Mometrix
TEST PREPARATION

ORELA®

Mathematics Secrets Study Guide Part 2 of 2

I0893669

Dear Future Exam Success Story

First of all, **THANK YOU** for purchasing Mometrix study materials!

Second, congratulations! You are one of the few determined test-takers who are committed to doing whatever it takes to excel on your exam. **You have come to the right place.** We developed these study materials with one goal in mind: to deliver you the information you need in a format that's concise and easy to use.

In addition to optimizing your guide for the content of the test, we've outlined our recommended steps for breaking down the preparation process into small, attainable goals so you can make sure you stay on track.

We've also analyzed the entire test-taking process, identifying the most common pitfalls and showing how you can overcome them and be ready for any curveball the test throws you.

Standardized testing is one of the biggest obstacles on your road to success, which only increases the importance of doing well in the high-pressure, high-stakes environment of test day. Your results on this test could have a significant impact on your future, and this guide provides the information and practical advice to help you achieve your full potential on test day.

Your success is our success

We would love to hear from you! If you would like to share the story of your exam success or if you have any questions or comments in regard to our products, please contact us at **800-673-8175** or **support@mometrix.com**.

Thanks again for your business and we wish you continued success!

Sincerely,
The Mometrix Test Preparation Team

TABLE OF CONTENTS

TRIGONOMETRY AND CALCULUS 1
- TRIGONOMETRIC FUNCTIONS 1
- COMPLEX NUMBERS 4
- THE UNIT CIRCLE 4
- DEFINED AND RECIPROCAL FUNCTIONS 6
- INVERSE TRIGONOMETRIC FUNCTIONS 6
- TRIGONOMETRIC IDENTITIES 7
- DOMAIN, RANGE, AND ASYMPTOTES IN TRIGONOMETRY 9
- RECTANGULAR AND POLAR COORDINATES 10
- CALCULUS 11
- CHAPTER QUIZ 33

STATISTICS, PROBABILITY, AND DISCRETE MATHEMATICS 34
- DISPLAYING DATA 34
- STATISTICS 41
- PROBABILITY 49
- SET THEORY 56
- CHAPTER QUIZ 64

ORELA PRACTICE TEST 65

ANSWER KEY AND EXPLANATIONS 94

HOW TO OVERCOME TEST ANXIETY 120

ONLINE RESOURCES 126

Trigonometry and Calculus

Trigonometric Functions

The basic trigonometric functions are sine (abbreviated 'sin'), cosine (abbreviated 'cos'), and tangent (abbreviated 'tan'). The simplest way to think of them is as describing the ratio of the side lengths of a right triangle in relation to the angles of the triangle.

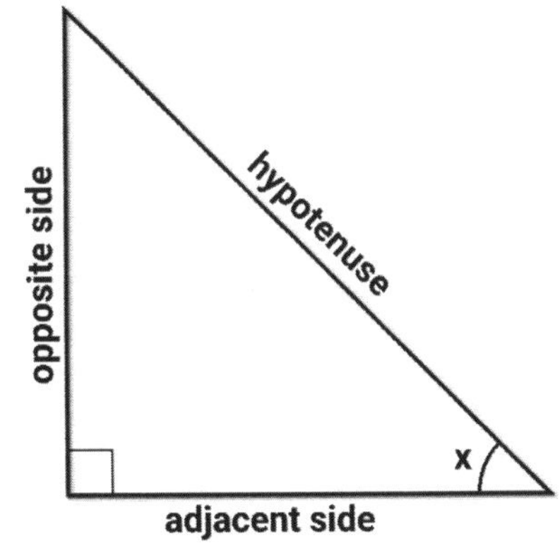

$$\sin x = \frac{\text{opposite side}}{\text{hypotenuse}}$$

$$\cos x = \frac{\text{adjacent side}}{\text{hypotenuse}}$$

$$\tan x = \frac{\text{opposite side}}{\text{adjacent side}}$$

Applying Trigonometric Functions

Suppose we have a point (x, y) at distance r from the origin such that the line segment between this point and the origin forms an angle of measure θ with the positive x-axis (We allow θ to be any real number, with the angle measured counterclockwise from the x-axis if $\theta > 0$ and clockwise if $\theta < 0$). Then we can define the trigonometric functions sine, cosine, and tangent by the formulas $\sin\theta = y/r$, $\cos\theta = x/r$, and $\tan\theta = y/x$ (and their reciprocal functions cosecant, secant, and cotangent accordingly). These functions are periodic because increasing the angle θ by 2π radians (that is, 360°) turns the line segment all the way around the origin so that it ends at the same point (x, y) again.

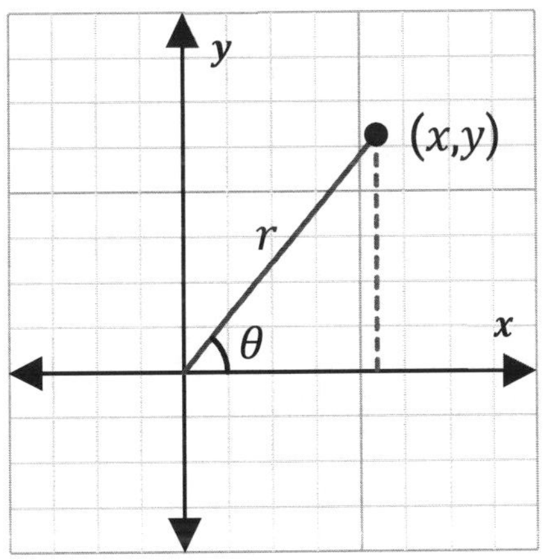

If the angle in the above diagram is acute, we can think of the line segment as the hypotenuse of a right triangle so that we get the familiar formulas $\sin\theta = (\text{opposite leg})/\text{hypotenuse}$, $\cos\theta = (\text{adjacent leg})/\text{hypotenuse}$, and $\tan\theta = (\text{opposite leg})/(\text{adjacent leg})$. In this form, the trigonometric functions are useful for finding missing sides of triangles. For example, suppose that standing 30 feet from the base of a tree, we have to look up at a 58° angle to see the top of the tree (see diagram). If we let x be the height of the tree, then we get the equation $\tan 58° = x/(30 \text{ feet})$. Multiplying both sides of the equation by 30 gives us the height of the tree, namely, $x = 30 \tan 58° \approx 48$ feet.

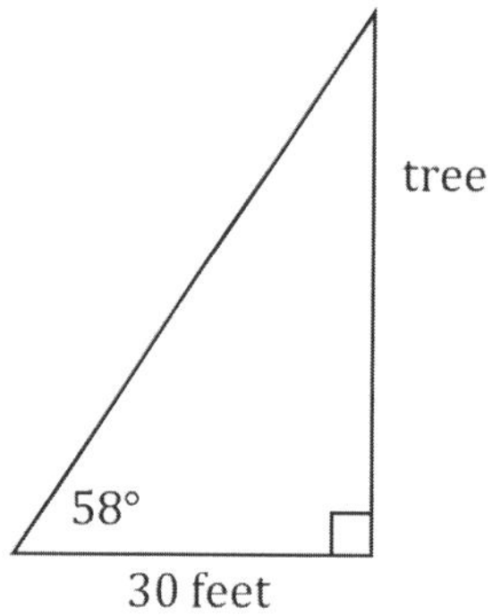

Graphing Trigonometric Functions

Sine

To graph the sine function, we let x (rather than θ) be the angle measure (usually in radians) and y be the value of the sine. That is, we graph the equation $y = \sin x$. From the definition of sine above we can show that $\sin 0 = 0$, $\sin(\pi/2) = 1$, $\sin \pi = 0$, $\sin(3\pi/2) = -1$, and $\sin 2\pi = 0$. Thus, the graph of the sine function begins at the origin, peaks at the point $(\pi/2, 1)$, crosses the x-axis at $(\pi, 0)$, has its lowest point at $(3\pi/2, -1)$, and returns to the x-axis at $(2\pi, 0)$. Because the sine function is periodic with period 2π radians (or 360°), this curve represents one complete cycle of the graph. The remainder of the graph of $y = \sin x$ consists of infinitely many repetitions of this cycle, producing the familiar "sine wave."

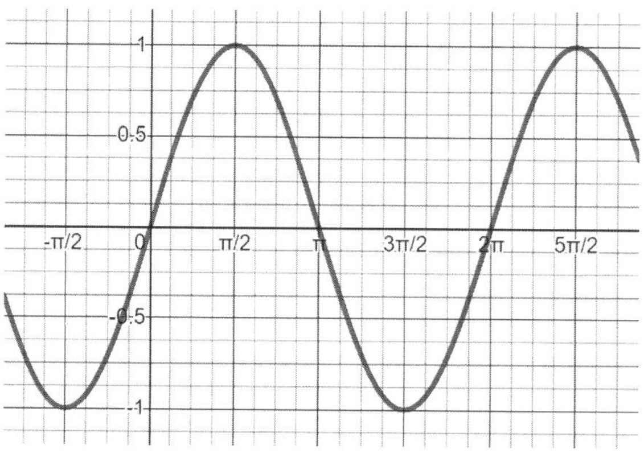

Review Video: Sine
Visit mometrix.com/academy and enter code: 193339

By suitable choice of the parameters A, B, C, and D, we can construct a "custom" sine function $y = A\sin(Bx + C) + D$ with any desired amplitude (the vertical distance the curve extends above and below its midline), period (distance between successive peaks), phase shift (horizontal shift), and midline (the horizontal line halfway between the peaks and troughs).

COSINE

Similarly, the graph of the cosine function is the graph of the equation $y = \cos x$. Proceeding as above, we get a graph with the same shape as that of the sine function but shifted by $\pi/2$ to the left.

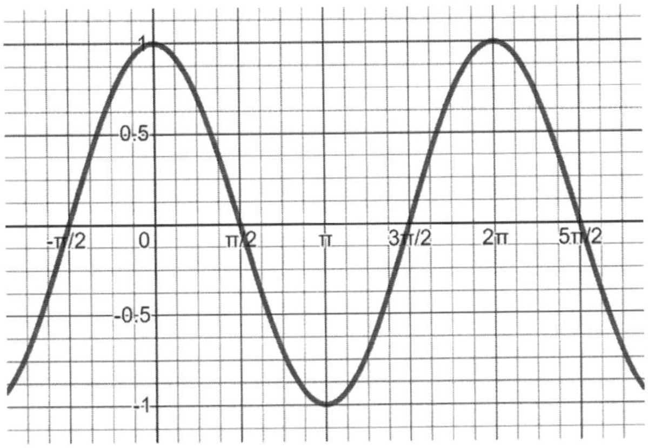

Review Video: Cosine
Visit mometrix.com/academy and enter code: 361120

As for the sine function, we can construct custom cosine graphs with desired amplitude, period, phase shift, and midline by choosing suitable values of the parameters in the equation $y = A\cos(Bx + C) + D$.

TANGENT

The graph of the tangent function (that is, the graph of the equation $y = \tan x$) turns out to be periodic with period π radians (or 180°). It consists of an infinite sequence of smooth curves crossing the x-axis at every multiple of π radians (180°) and having as asymptotes the vertical lines $x = \frac{k\pi}{2}$ radians ($x = k \times 90°$), for every odd integer k (which also means that the domain of the tangent function excludes these x-values).

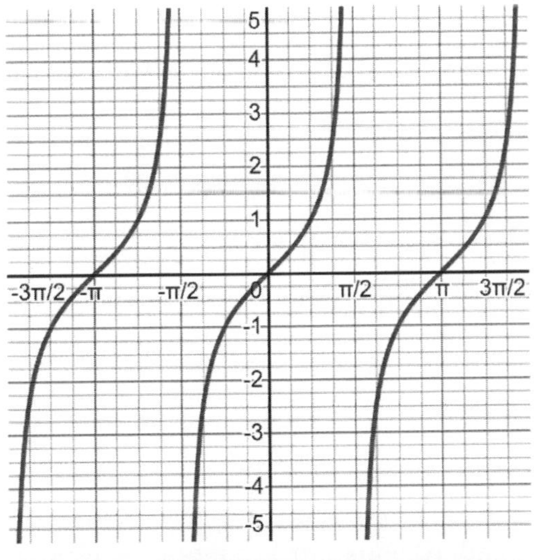

Review Video: Tangent
Visit mometrix.com/academy and enter code: 947639

Complex Numbers

Complex numbers consist of a real component and an imaginary component. Complex numbers are expressed in the form $a + bi$ with real component a and imaginary component bi. The imaginary unit i is equal to $\sqrt{-1}$. That means $i^2 = -1$. The imaginary unit provides a way to find the square root of a negative number. For example, $\sqrt{-25}$ is $5i$. You should expect questions asking you to add, subtract, multiply, divide, and simplify complex numbers. You may see a question that says, "Add $3 + 2i$ and $5 - 7i$" or "Subtract $4 + i\sqrt{5}$ from $2 + i\sqrt{5}$." Or you may see a question that says, "Multiply $6 + 2i$ by $8 - 4i$" or "Divide $1 - 3i$ by $9 - 7i$."

Operations on Complex Numbers

Operations with complex numbers resemble operations with variables in algebra. When adding or subtracting complex numbers, you can only combine like terms—real terms with real terms and imaginary terms with imaginary terms. For example, if you are asked to simplify the expression $-2 + 4i - (-3 + 7i) - 5i$, you should first remove the parentheses to yield $-2 + 4i + 3 - 7i - 5i$. Combining like terms yields $1 - 8i$. One interesting aspect of imaginary numbers is that if i has an exponent greater than 1, it can be simplified. Example: $i^2 = -1$, $i^3 = -i$, and $i^4 = 1$. When multiplying complex numbers, remember to simplify each i with an exponent greater than 1. For example, you might see a question that says, "Simplify $(2 - i)(3 + 2i)$." You need to distribute and multiply to get $6 + 4i - 3i - 2i^2$. This is further simplified to $6 + i - 2(-1)$, or $8 + i$.

Simplifying Expressions with Complex Denominators

If an expression contains an i in the denominator, it must be simplified. Remember, roots cannot be left in the denominator of a fraction. Since i is equivalent to $\sqrt{-1}$, i cannot be left in the denominator of a fraction. You must rationalize the denominator of a fraction that contains a complex denominator by multiplying the numerator and denominator by the conjugate of the denominator. The conjugate of the complex number $a + bi$ is $a - bi$. You can simplify $\frac{2}{5i}$ by simply multiplying $\frac{2}{5i} \times \frac{i}{i}$, which yields $-\frac{2}{5}i$. And you can simplify $\frac{5+3i}{2-4i}$ by multiplying $\frac{5+3i}{2-4i} \times \frac{2+4i}{2+4i}$. This yields $\frac{10+20i+6i-12}{4-8i+8i+16}$ which simplifies to $\frac{-2+26i}{20}$ or $\frac{-1+13i}{10}$, which can also be written as $-\frac{1}{10} + \frac{13}{10}i$.

The Unit Circle

Degrees, Radians, and the Unit Circle

It is important to understand the deep connection between trigonometry and circles. Specifically, the two main units, **degrees** (°) and **radians** (rad), that are used to measure angles are related this way: 360° in one full circle and 2π radians in one full circle: ($360° = 2\pi$ rad). The conversion factor relating the two is often stated as $\frac{180°}{\pi}$. For example, to convert $\frac{3\pi}{2}$ radians to degrees, multiply by the conversion factor: $\frac{3\pi}{2} \times \frac{180°}{\pi} = 270°$. As another example, to convert 60° to radians, divide by the conversion factor or multiply by the reciprocal: $60° \times \frac{\pi}{180°} = \frac{\pi}{3}$ radians.

Recall that the standard equation for a circle is $(x - h)^2 + (y - k)^2 = r^2$. A **unit circle** is a circle with a radius of 1 ($r = 1$) that has its center at the origin ($h = 0, k = 0$). Thus, the equation for the unit circle simplifies from the standard equation down to $x^2 + y^2 = 1$.

Standard position is the position of an angle of measure θ whose vertex is at the origin, the initial side crosses the unit circle at the point $(1, 0)$, and the terminal side crosses the unit circle at some other point (a, b). In the standard position, $\sin\theta = b$, $\cos\theta = a$, and $\tan\theta = \frac{b}{a}$.

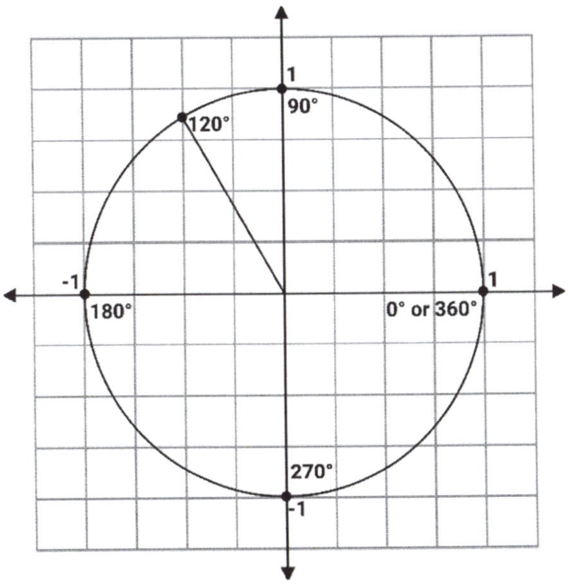

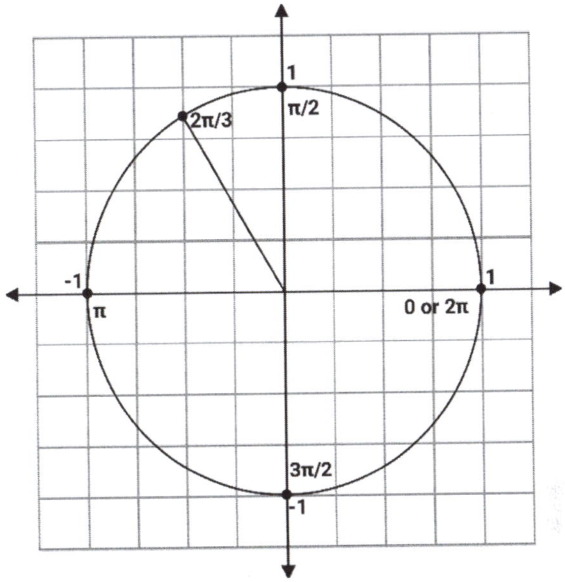

Review Video: Unit Circle
Visit mometrix.com/academy and enter code: 333922

TABLE OF COMMONLY ENCOUNTERED ANGLES

$0° = 0$ radians, $30° = \frac{\pi}{6}$ radians, $45° = \frac{\pi}{4}$ radians, $60° = \frac{\pi}{3}$ radians, and $90° = \frac{\pi}{2}$ radians

$\sin 0° = 0$	$\cos 0° = 1$	$\tan 0° = 0$
$\sin 30° = \frac{1}{2}$	$\cos 30° = \frac{\sqrt{3}}{2}$	$\tan 30° = \frac{\sqrt{3}}{3}$
$\sin 45° = \frac{\sqrt{2}}{2}$	$\cos 45° = \frac{\sqrt{2}}{2}$	$\tan 45° = 1$
$\sin 60° = \frac{\sqrt{3}}{2}$	$\cos 60° = \frac{1}{2}$	$\tan 60° = \sqrt{3}$
$\sin 90° = 1$	$\cos 90° = 0$	$\tan 90° =$ undefined
$\csc 0° =$ undefined	$\sec 0° = 1$	$\cot 0° =$ undefined
$\csc 30° = 2$	$\sec 30° = \frac{2\sqrt{3}}{3}$	$\cot 30° = \sqrt{3}$
$\csc 45° = \sqrt{2}$	$\sec 45° = \sqrt{2}$	$\cot 45° = 1$
$\csc 60° = \frac{2\sqrt{3}}{3}$	$\sec 60° = 2$	$\cot 60° = \frac{\sqrt{3}}{3}$
$\csc 90° = 1$	$\sec 90° =$ undefined	$\cot 90° = 0$

The values in the upper half of this table are values you should have memorized or be able to find quickly and those in the lower half can easily be determined as the reciprocal of the corresponding function.

Defined and Reciprocal Functions

The tangent function is defined as the ratio of the sine to the cosine: $\tan x = \frac{\sin x}{\cos x}$.

To take the reciprocal of a number means to place that number as the denominator of a fraction with a numerator of 1. The reciprocal functions are thus defined quite simply.

Cosecant	$\csc x$	$\frac{1}{\sin x}$
Secant	$\sec x$	$\frac{1}{\cos x}$
Cotangent	$\cot x$	$\frac{1}{\tan x}$

It is important to know these reciprocal functions, but they are not as commonly used as the three basic functions.

Review Video: Defined and Reciprocal Functions
Visit mometrix.com/academy and enter code: 996431

Inverse Trigonometric Functions

Inverse Functions

Each of the trigonometric functions accepts an angular measure, either degrees or radians, and gives a numerical value as the output. The inverse functions do the opposite; they accept a numerical value and give an angular measure as the output.

The inverse of sine, or arcsine, commonly written as either $\sin^{-1} x$ or $\arcsin x$, gives the angle whose sine is x. Similarly:

The inverse of $\cos x$ is written as $\cos^{-1} x$ or $\arccos x$ and means the angle whose cosine is x.
The inverse of $\tan x$ is written as $\tan^{-1} x$ or $\arctan x$ and means the angle whose tangent is x.
The inverse of $\csc x$ is written as $\csc^{-1} x$ or $\text{arccsc}\, x$ and means the angle whose cosecant is x.
The inverse of $\sec x$ is written as $\sec^{-1} x$ or $\text{arcsec}\, x$ and means the angle whose secant is x.

The inverse of $\cot x$ is written as $\cot^{-1} x$ or $\text{arccot}\, x$ and means the angle whose cotangent is x.

Review Video: Inverse Trig Functions
Visit mometrix.com/academy and enter code: 156054

Important Note About Solving Trigonometric Equations

When solving for an angle with a known trigonometric value, you must consider the sign and include all angles with that value. Your calculator will probably only give one value as an answer, typically in the following ranges:

- For $\sin^{-1} x$, $\left[-\frac{\pi}{2}, \frac{\pi}{2}\right]$ or $[-90°, 90°]$
- For $\cos^{-1} x$, $[0, \pi]$ or $[0°, 180°]$
- For $\tan^{-1} x$, $\left[-\frac{\pi}{2}, \frac{\pi}{2}\right]$ or $[-90°, 90°]$

It is important to determine if there is another angle in a different quadrant that also satisfies the problem. To do this, find the other quadrant(s) with the same sign for that trigonometric function and find the angle that has the same reference angle. Then check whether this angle is also a solution.

- In the first quadrant, all six trigonometric functions are positive.
- In the second quadrant, sin and csc are positive.
- In the third quadrant, tan and cot are positive.
- In the fourth quadrant, cos and sec are positive.

If you remember the phrase, "ALL Students Take Classes," you will be able to remember the sign of each trigonometric function in each quadrant. ALL represents all the signs in the first quadrant. The "S" in "Students" represents the sine function and its reciprocal in the second quadrant. The "T" in "Take" represents the tangent function and its reciprocal in the third quadrant. The "C" in "Classes" represents the cosine function and its reciprocal.

Trigonometric Identities

Sum and Difference

To find the sine, cosine, or tangent of the sum or difference of two angles, use one of the following formulas where α and β are two angles with known sine, cosine, or tangent values as needed:

$$\sin(\alpha \pm \beta) = \sin\alpha\cos\beta \pm \cos\alpha\sin\beta$$
$$\cos(\alpha \pm \beta) = \cos\alpha\cos\beta \mp \sin\alpha\sin\beta$$
$$\tan(\alpha \pm \beta) = \frac{\tan\alpha \pm \tan\beta}{1 \mp \tan\alpha\tan\beta}$$

Half Angle

To find the sine or cosine of half of a known angle, use the following formulas where θ is an angle with a known exact cosine value:

$$\sin\left(\frac{\theta}{2}\right) = \pm\sqrt{\frac{(1 - cos\theta)}{2}}$$

$$\cos\left(\frac{\theta}{2}\right) = \pm\sqrt{\frac{(1 + \cos\theta)}{2}}$$

To determine the sign of the answer, you must recognize which quadrant the given angle is in and apply the correct sign for the trigonometric function you are using. If you need to find an expression for the exact sine or cosine of an angle that you do not know, such as sine 22.5°, you can rewrite the given angle as a half angle, such as $\sin\left(\frac{45°}{2}\right)$, and use the formula above:

$$\sin\left(\frac{45°}{2}\right) = \pm\sqrt{\frac{(1 - \cos(45°))}{2}} = \pm\sqrt{\frac{\left(1 - \frac{\sqrt{2}}{2}\right)}{2}} = \pm\sqrt{\frac{(2 - \sqrt{2})}{4}} = \pm\frac{1}{2}\sqrt{(2 - \sqrt{2})}$$

To find the tangent or cotangent of half of a known angle, use the following formulas where θ is an angle with known exact sine and cosine values:

$$\tan\frac{\theta}{2} = \frac{sin\theta}{1 + \cos\theta}$$

$$\cot\frac{\theta}{2} = \frac{\sin\theta}{1 - \cos\theta}$$

These formulas will work for finding the tangent or cotangent of half of any angle unless the cosine of θ happens to make the denominator of the identity equal to 0.

The Pythagorean theorem states that $a^2 + b^2 = c^2$ for all right triangles. The trigonometric identity that derives from this principle is stated in this way: $\sin^2\theta + \cos^2\theta = 1$.

Dividing each term by either $\sin^2\theta$ or $\cos^2\theta$ yields two other identities, respectively:

$$1 + \cot^2\theta = \csc^2\theta$$

$$\tan^2\theta + 1 = \sec^2\theta$$

Review Video: Sum and Difference Trigonometric Identities
Visit mometrix.com/academy and enter code: 468838

DOUBLE ANGLES

In each case, use one of the double angle formulas. To find the sine or cosine of twice a known angle, use one of the following formulas:

$$\sin(2\theta) = 2\sin\theta\cos\theta$$

$$\begin{aligned}\cos(2\theta) &= \cos^2\theta - \sin^2\theta \\ &= 2\cos^2\theta - 1 \\ &= 1 - 2\sin^2\theta\end{aligned}$$

To find the tangent or cotangent of twice a known angle, use the formulas where θ is an angle with known exact sine, cosine, tangent, and cotangent values:

$$\tan(2\theta) = \frac{2\tan\theta}{1 - \tan^2\theta}$$

$$\cot(2\theta) = \frac{\cot\theta - \tan\theta}{2}$$

PRODUCTS

To find the product of the sines and cosines of two different angles, use one of the following formulas where α and β are two unique angles:

$$\sin\alpha\sin\beta = \frac{1}{2}[\cos(\alpha-\beta) - \cos(\alpha+\beta)]$$

$$\cos\alpha\cos\beta = \frac{1}{2}[\cos(\alpha+\beta) + \cos(\alpha-\beta)]$$

$$\sin\alpha\cos\beta = \frac{1}{2}[\sin(\alpha+\beta) + \sin(\alpha-\beta)]$$

$$\cos\alpha\sin\beta = \frac{1}{2}[\sin(\alpha+\beta) - \sin(\alpha-\beta)]$$

Review Video: Half-Angle, Double Angle, and Product Trig Identities
Visit mometrix.com/academy and enter code: 274252

COMPLEMENTARY

The trigonometric cofunction identities use the trigonometric relationships of complementary angles (angles whose sum is 90°). These are:

$$\cos x = \sin(90° - x)$$

$$\csc x = \sec(90° - x)$$

$$\cot x = \tan(90° - x)$$

Domain, Range, and Asymptotes in Trigonometry

The domain is the set of all possible real number values of x on the graph of a trigonometric function. Some graphs will impose limits on the values of x.

The range is the set of all possible real number values of y on the graph of a trigonometric function. Some graphs will impose limits on the values of y.

Asymptotes are lines that the graph of a trigonometric function approaches but never reaches. Asymptotes exist for values of x in the graphs of the tangent, cotangent, secant, and cosecant. The sine and cosine graphs do not have any asymptotes.

Domain, Range, and Asymptotes of the Six Trigonometric Functions

The domain, range, and asymptotes for each of the trigonometric functions are as follows:

- In the **sine** function, the domain is all real numbers, the range is $-1 \leq y \leq 1$, and there are no asymptotes.
- In the **cosine** function, the domain is all real numbers, the range is $-1 \leq y \leq 1$, and there are no asymptotes.
- In the **tangent** function, the domain is $x \in \mathbb{R}; x \neq \frac{\pi}{2} + k\pi$, the range is all real numbers, and the asymptotes are the lines $x = \frac{\pi}{2} + k\pi$.
- In the **cosecant** function, the domain is $x \in \mathbb{R}; x \neq k\pi$, the range is $(-\infty, -1]$ and $[1, \infty)$, and the asymptotes are the lines $x = k\pi$.
- In the **secant** function, the domain is $x \in \mathbb{R}; x \neq \frac{\pi}{2} + k\pi$, the range is $(-\infty, 1]$ and $[1, \infty)$, and the asymptotes are the lines $x = \frac{\pi}{2} + k\pi$.
- In the **cotangent** function, the domain is $x \in \mathbb{R}; x \neq k\pi$, the range is all real numbers, and the asymptotes are the lines $x = k\pi$.

In each of the above cases, k represents any integer.

Rectangular and Polar Coordinates

Rectangular coordinates are those that lie on the square grids of the Cartesian plane. They should be quite familiar to you. The polar coordinate system is based on a circular graph, rather than the square grid of the Cartesian system. Points in the polar coordinate system are in the format (r, θ), where r is the distance from the origin (think radius of the circle) and θ is the smallest positive angle (moving counterclockwise around the circle) made with the positive horizontal axis.

To convert a point from rectangular (x, y) format to polar (r, θ) format, use the formula (x, y) to $(r, \theta) \Rightarrow r = \sqrt{x^2 + y^2}; \theta = \arctan\frac{y}{x}$ when $x \neq 0$.

If x is positive, use the positive square root value for r. If x is negative, use the negative square root value for r. If $x = 0$, use the following rules:

- If $y = 0$, then $\theta = 0$.
- If $y > 0$, then $\theta = \frac{\pi}{2}$.
- If $y < 0$, then $\theta = \frac{3\pi}{2}$.

To convert a point from polar (r, θ) format to rectangular (x, y) format, use the formula (r, θ) to $(x, y) \Rightarrow x = r\cos\theta\ ; y = r\sin\theta$.

De Moivre's Theorem

De Moivre's theorem is used to find the powers of complex numbers (numbers that contain the imaginary number i) written in polar form. Given a trigonometric expression that contains i, such as $z = r\cos x + ir\sin x$, where r is a real number and x is an angle measurement in polar form, use the formula $z^n = r^n(\cos nx + i\sin nx)$, where r and n are real numbers, x is the angle measure in polar form, and i is the imaginary number $i = \sqrt{-1}$. The expression $\cos x + i\sin x$ can be written $\text{cis}\, x$, making the formula appear in the format $z^n = r^n \text{cis}\, nx$.

Note that De Moivre's theorem is only for angles in polar form. If you are given an angle in degrees, you must convert to polar form before using the formula.

Calculus

Calculus, also called analysis, is the branch of mathematics that studies the length, area, and volume of objects, and the rate of change of quantities (which can be expressed as slopes of curves). The two principal branches of calculus are differential and integral. **Differential calculus** is based on derivatives and takes the form,

$$\frac{d}{dx}f(x)$$

Integral calculus is based on integrals and takes the form,

$$\int f(x)dx$$

Some of the basic ideas of calculus were utilized as far back in history as Archimedes. However, its modern forms were developed by Newton and Leibniz.

Limits

The **limit of a function** is represented by the notation $\lim_{x\to a} f(x)$. It is read as "the limit of f of x as x approaches a." In many cases, $\lim_{x\to a} f(x)$ will simply be equal to $f(a)$, but not always. Limits are important because some functions are not defined or are not easy to evaluate at certain values of x.

The limit at the point is said to exist only if the limit is the same when approached from the right side as from the left: $\lim_{x\to a^+} f(x) = \lim_{x\to a^-} f(x)$). Notice the symbol by the a in each case. When x approaches a from the right, it approaches from the positive end of the number line. When x approaches a from the left, it approaches from the negative end of the number line.

If the limit as x approaches a differs depending on the direction from which it approaches, then the limit does not exist at a. In other words, if $\lim_{x\to a^+} f(x)$ does not equal $\lim_{x\to a^-} f(x)$, then the limit does not exist at a. The limit also does not exist if either of the one-sided limits does not exist.

Situations in which the limit does not exist include a function that jumps from one value to another at a, one that oscillates between two different values as x approaches a, or one that increases or decreases without bounds as x approaches a. If the limit you calculate has a value of $\frac{c}{0}$, where c is any constant, this means the function goes to infinity and the limit does not exist.

It is possible for two functions that do not have limits to be multiplied to get a new function that does have a limit. Just because two functions do not have limits, do not assume that the product will not have a limit.

Direct Substitution

The first thing to try when looking for a limit is direct substitution. To find the limit of a function $\lim_{x \to a} f(x)$ by direct substitution, substitute the value of a for x in the function and solve. The following patterns apply to finding the limit of a function by direct substitution:

$$\lim_{x \to a} b = b, \text{ where } b \text{ is any real number}$$

$$\lim_{x \to a} x = a$$

$$\lim_{x \to a} x^n = a^n, \text{ where } n \text{ is any positive integer}$$

$$\lim_{x \to a} \sqrt{x} = \sqrt{a}; a > 0$$

$$\lim_{x \to a} \sqrt[n]{x} = \sqrt[n]{a}, \text{ where } n \text{ is a positive integer and } a > 0 \text{ for all even values of } n$$

$$\lim_{x \to a} \frac{1}{x} = \frac{1}{a}; a \neq 0$$

You can also use substitution for finding the limit of a trigonometric function, a polynomial function, or a rational function. Be sure that in manipulating an expression to find a limit that you do not divide by terms equal to zero.

In finding the limit of a composite function, begin by finding the limit of the innermost function. For example, to find $\lim_{x \to a} f(g(x))$, first find the value of $\lim_{x \to a} g(x)$. Then substitute this value for x in $f(x)$ and solve. The result is the limit of the original problem.

Limits and Operations

When finding the limit of the sum or difference of two functions, find the limit of each individual function and then add or subtract the results. Example:

$$\lim_{x \to a}[f(x) \pm g(x)] = \lim_{x \to a} f(x) \pm \lim_{x \to a} g(x)$$

To find the limit of the product or quotient of two functions, find the limit of each individual function and then multiply or divide the results. Example:

$$\lim_{x \to a}[f(x) \times g(x)] = \lim_{x \to a} f(x) \times \lim_{x \to a} g(x)$$

$$\lim_{x \to a} \frac{f(x)}{g(x)} = \frac{\lim_{x \to a} f(x)}{\lim_{x \to a} g(x)}, \text{ where } g(x) \neq 0$$

$$\lim_{x \to a} g(x) \neq 0$$

When finding the quotient of the limits of two functions, make sure the denominator is not equal to zero. If it is, use differentiation or L'Hôpital's rule to find the limit.

To find the limit of a power of a function or a root of a function, find the limit of the function and then raise the limit to the original power or take the root of the limit. Example:

$$\lim_{x\to a}[f(x)]^n = \left[\lim_{x\to a} f(x)\right]^n$$

$$\lim_{x\to a} \sqrt[n]{f(x)} = \sqrt[n]{\lim_{x\to a} f(x)}, \text{ where } n \text{ is a positive integer}$$

$$\lim_{x\to a} f(x) > 0 \text{ for all even values of } n$$

To find the limit of a function multiplied by a scalar, find the limit of the function and multiply the result by the scalar. Example:

$$\lim_{x\to a} kf(x) = k \lim_{x\to a} f(x), \text{ where } k \text{ is a real number.}$$

Review Video: Limits
Visit mometrix.com/academy and enter code: 554961

L'Hôpital's Rule

Sometimes solving $\lim_{x\to a}\frac{f(x)}{g(x)}$ by the direct substitution method will result in the numerator and denominator both being equal to zero, or both being equal to infinity. This outcome is called an indeterminate form. The limit cannot be directly found by substitution in these cases. L'Hôpital's rule is a useful method for finding the limit of a problem in the indeterminate form. L'Hôpital's rule allows you to find the limit using derivatives. Assuming both the numerator and denominator are differentiable, and that both are equal to zero when the direct substitution method is used, take the derivative of both the numerator and the denominator and then use the direct substitution method. For example, if $\lim_{x\to a}\frac{f(x)}{g(x)} = \frac{0}{0}$, take the derivatives of $f(x)$ and $g(x)$ and then find $\lim_{x\to a}\frac{f'(x)}{g'(x)}$. If $g'(x) \neq 0$, then you have found the limit of the original function. If $g'(x) = 0$ and $f'(x) = 0$, L'Hôpital's rule may be applied to the function $\frac{f'(x)}{g'(x)}$, and so on until either a limit is found, or it can be determined that the limit does not exist.

Review Video: L'Hopital's Rule
Visit mometrix.com/academy and enter code: 624400

Squeeze Theorem

The squeeze theorem is known by many names, including the sandwich theorem, the sandwich rule, the squeeze lemma, the squeezing theorem, and the pinching theorem. No matter what you call it, the principle is the same. To prove the limit of a difficult function exists, find the limits of two functions, one on either side of the unknown, that are easy to compute. If the limits of these functions are equal, then that is also the limit of the unknown function. In mathematical terms, the theorem is:

If $g(x) \leq f(x) \leq h(x)$ for all values of x where $f(x)$ is the function with the unknown limit, and if $\lim_{x \to a} g(x) = \lim_{x \to a} h(x)$, then this limit is also equal to $\lim_{x \to a} f(x)$.

To find the limit of an expression containing an absolute value sign, take the absolute value of the limit. If $\lim_{n \to \infty} a_n = L$, where L is the numerical value for the limit, then $\lim_{n \to \infty} |a_n| = |L|$. Also, if $\lim_{n \to \infty} |a_n| = 0$, then $\lim_{n \to \infty} a_n = 0$. The trick comes when you are asked to find the limit as n approaches from the left. Whenever the limit is being approached from the left, it is being approached from the negative end of the domain. The absolute value sign makes everything in the equation positive, essentially eliminating the negative side of the domain. In this case, rewrite the equation without the absolute value signs and add a negative sign in front of the expression. Example:

$$\lim_{n \to 0^-} |x| \text{ becomes } \lim_{n \to 0^-} (-x)$$

Review Video: Squeeze Theorem
Visit mometrix.com/academy and enter code: 383104

Derivatives

The derivative of a function is a measure of how much that function is changing at a specific point, and is the slope of a line tangent to a curve at the specific point. The derivative of a function $f(x)$ is written $f'(x)$, and read, "f prime of x." Other notations for the derivative include $D_x f(x)$, y', $D_x y$, $\frac{dy}{dx}$, and $\frac{d}{dx} f(x)$. The definition of the derivative of a function is $f'(x) = \lim_{h \to 0} \frac{f(x+h)-f(x)}{h}$. However, this formula is rarely used.

There is a simpler method you can use to find the derivative of a polynomial. Given a function $f(x) = a_n x^n + a_{n-1} x^{n-1} + a_{n-2} x^{n-2} + \cdots + a_1 x + a_0$, multiply each exponent by its corresponding coefficient to get the new coefficient and reduce the value of the exponent by one. Coefficients with no variable are dropped. This gives $f'(x) = n a_n x^{n-1} + (n-1) a_{n-1} x^{n-2} + \cdots + a_1$, a pattern that can be repeated for each successive derivative.

Review Video: Definition of a Derivative
Visit mometrix.com/academy and enter code: 787269

Differentiable functions are functions that have a derivative. Some basic rules for finding derivatives of functions are:

$$f(x) = c \Rightarrow f'(x) = 0; \text{ where } c \text{ is a constant}$$
$$f(x) = x \Rightarrow f'(x) = 1$$
$$f(x) = x^n \Rightarrow f'(x) = nx^{n-1}; \text{ where } n \text{ is a real number}$$

$$\left(cf(x)\right)' = cf'(x); \text{ where } c \text{ is a constant}$$
$$(f+g)'(x) = f'(x) + g'(x)$$
$$(fg)'(x) = f(x)g'(x) + f'(x)g(x)$$
$$\left(\frac{f}{g}\right)'(x) = \frac{f'(x)g(x) - f(x)g'(x)}{[g(x)]^2}$$
$$(f \circ g)'(x) = f'\left(g(x)\right) \times g'(x)$$

This last formula is also known as the **chain rule**. If you are finding the derivative of a polynomial that is raised to a power, let the polynomial be represented by $g(x)$ and use the chain rule. The chain rule is one of the most important concepts to grasp in the early stages of learning calculus. Many other rules and shortcuts are based upon the chain rule.

Review Video: Derivative Properties and Formulas
Visit mometrix.com/academy and enter code: 735227

Review Video: Product and Quotient Rules in Calculus
Visit mometrix.com/academy and enter code: 649197

Review Video: The Chain Rule - An Integral Part of Calculus
Visit mometrix.com/academy and enter code: 938732

Difference Quotient and Derivative

A secant is a line that connects two points on a curve. The **difference quotient** gives the slope of an arbitrary secant line that connects the point $\left(x, f(x)\right)$ with a nearby point $\left(x+h, f(x+h)\right)$ on the graph of the function f. The difference quotient is the same formula that is always used to determine a slope—the change in y divided by the change in x. It is written as $\frac{f(x+h)-f(x)}{h}$.

A tangent is a line that touches a curve at one point. The tangent and the curve have the same slope at the point where they touch. The derivative is the function that gives the slope of both the tangent and the curve of the function at that point. The derivative is written as the limit of the difference quotient, or:

$$\lim_{h\to 0} \frac{f(x+h) - f(x)}{h}$$

If the function is f, the derivative is denoted as $f'(x)$, and it is the slope of the function f at point $\left(x, f(x)\right)$. It is expressed as:

$$f'(x) = \lim_{h\to 0} \frac{f(x+h) - f(x)}{h}$$

Implicit Functions

An **implicit function** is one where it is impossible, or very difficult, to express one variable in terms of another by normal algebraic methods. This would include functions that have both variables raised to a power greater than 1, functions that have two variables multiplied by each other, or a combination of the two. To differentiate such a function with respect to x, take the derivative of each term that contains a variable, either x or y. When differentiating a term with y, use the chain rule, first taking the derivative with respect to y, and then multiplying by $\frac{dy}{dx}$. If a term contains both

x and y, you will have to use the product rule as well as the chain rule. Once the derivative of each individual term has been found, use the rules of algebra to solve for $\frac{dy}{dx}$ to get the final answer.

Review Video: Implicit Differentiation
Visit mometrix.com/academy and enter code: 102151

Derivatives of Trigonometric Functions

Trigonometric functions are any functions that include one of the six trigonometric expressions. The following rules for derivatives apply for all trigonometric differentiation:

$$\frac{d}{dx}(\sin x) = \cos x, \qquad \frac{d}{dx}(\cos x) = -\sin x, \qquad \frac{d}{dx}(\tan x) = \sec^2 x$$

For functions that are a combination of trigonometric and algebraic expressions, use the chain rule:

$$\frac{d}{dx}(\sin u) = \cos u\frac{du}{dx} \qquad \frac{d}{dx}(\sec u) = \tan u \sec u\frac{du}{dx}$$
$$\frac{d}{dx}(\cos u) = -\sin u\frac{du}{dx} \qquad \frac{d}{dx}(\csc u) = -\csc u \cot u\frac{du}{dx}$$
$$\frac{d}{dx}(\tan u) = \sec^2 u\frac{du}{dx} \qquad \frac{d}{dx}(\cot u) = -\csc^2 u\frac{du}{dx}$$

Functions involving the inverses of the trigonometric functions can also be differentiated.

$$\frac{d}{dx}(\sin^{-1} u) = \frac{1}{\sqrt{1-u^2}}\frac{du}{dx} \qquad \frac{d}{dx}(\csc^{-1} u) = \frac{-1}{|u|\sqrt{u^2-1}}\frac{du}{dx}$$
$$\frac{d}{dx}(\cos^{-1} u) = \frac{-1}{\sqrt{1-u^2}}\frac{du}{dx} \qquad \frac{d}{dx}(\sec^{-1} u) = \frac{1}{|u|\sqrt{u^2-1}}\frac{du}{dx}$$
$$\frac{d}{dx}(\tan^{-1} u) = \frac{1}{1+u^2}\frac{du}{dx} \qquad \frac{d}{dx}(\cot^{-1} u) = \frac{-1}{1+u^2}\frac{du}{dx}$$

In each of the above expressions, u represents a differentiable function. Also, the value of u must be such that the radicand, if applicable, is a positive number. Remember the expression $\frac{du}{dx}$ means to take the derivative of the function u with respect to the variable x.

Review Video: Derivatives of Trigonometry Functions
Visit mometrix.com/academy and enter code: 132724

Derivatives of Exponential and Logarithmic Functions

Exponential functions are in the form e^x, which has itself as its derivative: $\frac{d}{dx}e^x = e^x$. For functions that have a function as the exponent rather than just an x, use the formula $\frac{d}{dx}e^u = e^u\frac{du}{dx}$. The inverse of the exponential function is the natural logarithm. To find the derivative of the natural logarithm, use the formula $\frac{d}{dx}\ln u = \frac{1}{u}\frac{du}{dx}$.

If you are trying to solve an expression with a variable in the exponent, use the formula $a^x = e^{x \ln a}$, where a is a positive real number and x is any real number. To find the derivative of a function in this format, use the formula $\frac{d}{dx}a^x = a^x \ln a$. If the exponent is a function rather than a single variable x, use the formula $\frac{d}{dx}a^u = a^u \ln a \frac{du}{dx}$. If you are trying to solve an expression involving a logarithm, use the formula $\frac{d}{dx}(\log_a x) = \frac{1}{x \ln a}$ or $\frac{d}{dx}(\log_a |u|) = \frac{1}{u \ln a}\frac{du}{dx}; u \neq 0$.

Review Video: Derivatives of Exponential and Logarithmic Functions
Visit mometrix.com/academy and enter code: 594367

Continuity

A function can be either continuous or discontinuous. A conceptual way to describe continuity is this: A function is continuous if its graph can be traced with a pen without lifting the pen from the page. In other words, there are no breaks or gaps in the graph of the function. However, this is only a description, not a technical definition. A function is continuous at the point $x = a$ if the three following conditions are met:

1. $f(a)$ is defined
2. $\lim_{x \to a} f(x)$ exists
3. $\lim_{x \to a} f(x) = f(a)$

If any of these conditions are not met, the function is discontinuous at the point $x = a$.

A function can be continuous at a point, continuous over an interval, or continuous everywhere. The above rules define continuity at a point. A function that is continuous over an interval $[a, b]$ is continuous at the points a and b and at every point between them. A function that is continuous everywhere is continuous for every real number, that is, for all points in its domain.

Discontinuity

Discontinuous functions are categorized according to the type or cause of discontinuity. Three examples are point, infinite, and jump discontinuity. A function with a point discontinuity has one value of x for which it is not continuous. A function with infinite discontinuity has a vertical asymptote at $x = a$ and $f(a)$ is undefined. It is said to have an infinite discontinuity at $x = a$. A function with jump discontinuity has one-sided limits from the left and from the right, but they are not equal to one another, that is, $\lim_{x \to a^-} f(x) \neq \lim_{x \to a^+} f(x)$. It is said to have a jump discontinuity at $x = a$.

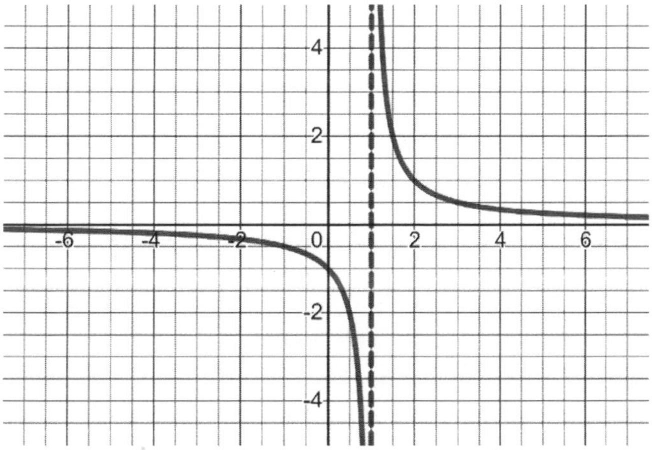

The function $f(x) = \sqrt{x}$ is not differentiable because its domain is $[0, \infty)$ and it has a discontinuity at $x = 0$. Therefore, a tangent could not be drawn at that point.

Differentiability

A function is said to be differentiable at point $x = a$ if it has a derivative at that point, that is, if $f'(a)$ exists. For a function to be differentiable, it must be continuous because the slope cannot be defined at a point of discontinuity. Furthermore, for a function to be differentiable, its graph must not have any sharp turn for which it is impossible to draw a tangent line. The sine function is an example of a differentiable function. It is continuous, and a tangent line can be drawn anywhere along its graph.

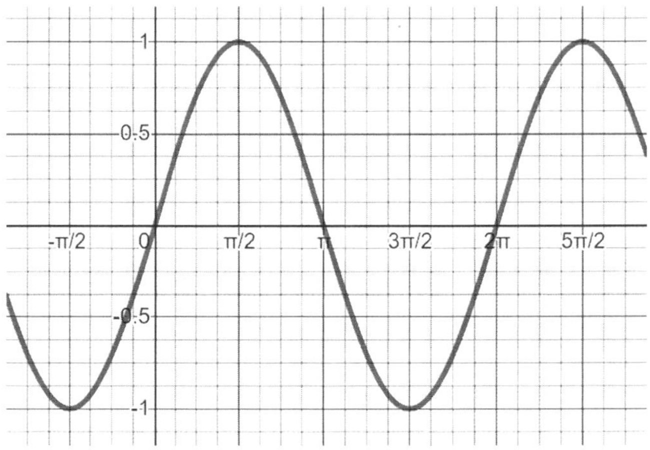

The absolute value function, $f(x) = |x|$, is an example of a function that is not differentiable:

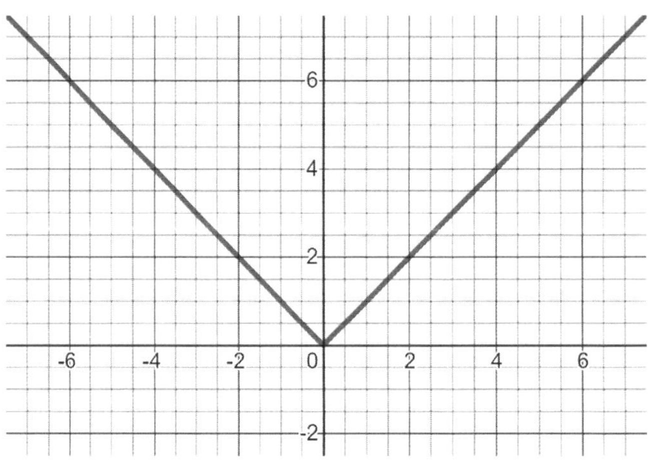

It is continuous, but it has a sharp turn at $x = 0$ which prohibits the drawing of a tangent at that point. All differentiable functions are continuous, but not all continuous functions are differentiable, as the absolute value function demonstrates.

The function $f(x) = \sqrt{x}$ is not differentiable because its domain is $[0, \infty)$ and it has a discontinuity at $x = 0$. Therefore, a tangent could not be drawn at that point.

Approximating a Derivative from a Table of Values

The derivative of a function at a particular point is equal to the slope of the graph of the function at that point. For a nonlinear function, it can be thought of as the limit of the slope of a line drawn between two other points on the function as those points become closer to the point in question. Such a line drawn through two points on the function is called a **secant** of the function.

This definition of the derivative in terms of the secant allows us to approximate the derivative of a function at a point from a table of values: we take the slope of the line through the points on either side. That is, if the point lies between (x_1, y_1) and (x_2, y_2), the slope of the secant—the approximate derivative—is $\frac{y_2-y_1}{x_2-x_1}$. (This is also equal to the average slope over the interval $[x_1, x_2]$.)

For example, consider the function represented by the following table:

x	0	2	4	6	8	10
y	1	5	8	9	7	4

Suppose we want to know the derivative of the function when $x = 3$. This lies between the points $(2, 5)$ and $(4, 8)$; the approximate derivative is $\frac{8-5}{4-2} = \frac{3}{2}$.

Position, Velocity, and Acceleration

Velocity is a specific type of rate of change. It refers to the rate of change of the position of an object with relation to a reference frame. **Acceleration** is the rate of change of velocity.

Average velocity over a period of time is found using the formula $\bar{v} = \frac{s(t_2)-s(t_1)}{t_2-t_1}$, where t_1 and t_2 are specific points in time and $s(t_1)$ and $s(t_2)$ are the distances traveled at those points in time.

Instantaneous velocity at a specific time, t, is found using the limit $v = \lim_{h\to 0} \frac{s(t+h)-s(t)}{h}$, or $v = s'(t)$.

Remember that velocity at a given point is found using the first derivative, and acceleration at a given point is found using the second derivative. Therefore, the formula for acceleration at a given point in time is found using the formula $a(t) = v'(t) = s''(t)$, where a is acceleration, v is velocity, and s is displacement.

Review Video: Position, Velocity, and Acceleration
Visit mometrix.com/academy and enter code: 714040

Using First and Second Derivatives

The **first derivative** of a function is equal to the **rate of change** of the function. The sign of the rate of change shows whether the value of the function is **increasing** or **decreasing**. A positive rate of change—and therefore a positive first derivative—represents that the function is increasing at that point. A negative rate of change represents that the function is decreasing. If the rate of change is zero, the function is not changing, i.e., it is constant.

For example, consider the function $f(x) = x^3 - 6x^2 - 15x + 12$. The derivative of this function is $f'(x) = 3x^2 - 12x - 15 = 3(x^2 - 4x - 5) = 3(x - 5)(x + 1)$. This derivative is a quadratic function with zeroes at $x = 5$ and $x = -1$; by plugging in points in each interval we can find that $f'(x)$ is positive when $x < -1$ and when $x > 5$ and negative when $-1 < x < 5$. Thus $f(x)$ is increasing in the interval $(-\infty, -1) \cup (5, \infty)$ and decreasing in the interval $(-1, 5)$.

Review Video: Rate of Change
Visit mometrix.com/academy and enter code: 100367

EXTREMA

The **maximum** and **minimum** values of a function are collectively called the **extrema** of the function. Both maxima and minima can be local, also known as relative, or absolute. A local maximum or minimum refers to the value of a function near a certain value of x. An absolute maximum or minimum refers to the value of a function on a given interval.

The local maximum of a function is the largest value that the function attains near a certain value of x. For example, function f has a local maximum at $x = b$ if $f(b)$ is the largest value that f attains as it approaches b.

Conversely, the local minimum is the smallest value that the function attains near a certain value of x. In other words, function f has a local minimum at $x = b$ if $f(b)$ is the smallest value that f attains as it approaches b.

The absolute maximum of a function is the largest value of the function over a certain interval. The function f has an absolute maximum at $x = b$ if $f(b) \geq f(x)$ for all x in the domain of f.

The absolute minimum of a function is the smallest value of the function over a certain interval. The function f has an absolute minimum at $x = b$ if $f(b) \leq f(x)$ for all x in the domain of f.

CRITICAL POINTS

Remember Rolle's theorem, which states that if two points have the same value in the range that there must be a point between them where the slope of the graph is zero. This point is located at a peak or valley on the graph. A **peak** is a maximum point, and a **valley** is a minimum point. The relative minimum is the lowest point on a graph for a given section of the graph. It may or may not be the same as the absolute minimum, which is the lowest point on the entire graph. The relative maximum is the highest point on one section of the graph. Again, it may or may not be the same as the absolute maximum. A relative extremum (plural extrema) is a relative minimum or relative maximum point on a graph.

A **critical point** is a point $(x, f(x))$ that is part of the domain of a function, such that either $f'(x) = 0$ or $f'(x)$ does not exist. If either of these conditions is true, then x is either an inflection point or a point at which the slope of the curve changes sign. If the slope changes sign, then a relative minimum or maximum occurs.

In graphing an equation with relative extrema, use a sign diagram to approximate the shape of the graph. Once you have determined the relative extrema, calculate the sign of a point on either side of each critical point. This will give a general shape of the graph, and you will know whether each critical point is a relative minimum, a relative maximum, or a point of inflection.

FIRST DERIVATIVE TEST

Remember that critical points occur where the slope of the curve is 0. Also remember that the **first derivative** of a function gives the slope of the curve at a particular point on the curve. Because of this property of the first derivative, the first derivative test can be used to determine if a critical point is a minimum or maximum. If $f'(x)$ is negative at a point to the left of a critical number and $f'(x)$ is positive at a point to the right of a critical number, then the critical number is a relative minimum. If $f'(x)$ is positive to the left of a critical number and $f'(x)$ is negative to the right of a critical number, then the critical number is a relative maximum. If $f'(x)$ has the same sign on both sides, then the critical number is a point of inflection.

Review Video: First Derivative Test
Visit mometrix.com/academy and enter code: 205981

Second Derivative Test

The **second derivative**, designated by $f''(x)$, is helpful in determining whether the relative extrema of a function are relative maximums or relative minimums. If the second derivative at the critical point is greater than zero, the critical point is a relative minimum. If the second derivative at the critical point is less than zero, the critical point is a relative maximum. If the second derivative at the critical point is equal to zero, you must use the first derivative test to determine whether the point is a relative minimum or a relative maximum.

There are a couple of ways to determine the concavity of the graph of a function. To test a portion of the graph that contains a point with domain p, find the second derivative of the function and evaluate it for p. If $f''(p) > 0$, then the graph is concave upward at that point. If $f''(p) < 0$, then the graph is concave downward at that point.

The **point of inflection** on the graph of a function is the point at which the concavity changes from concave downward to concave upward or from concave upward to concave downward. The easiest way to find the points of inflection is to find the second derivative of the function and then solve the equation $f''(x) = 0$. Remember that if $f''(p) > 0$, the graph is concave upward, and if $f''(p) < 0$, the graph is concave downward. Logically, the concavity changes at the point when $f''(p) = 0$:

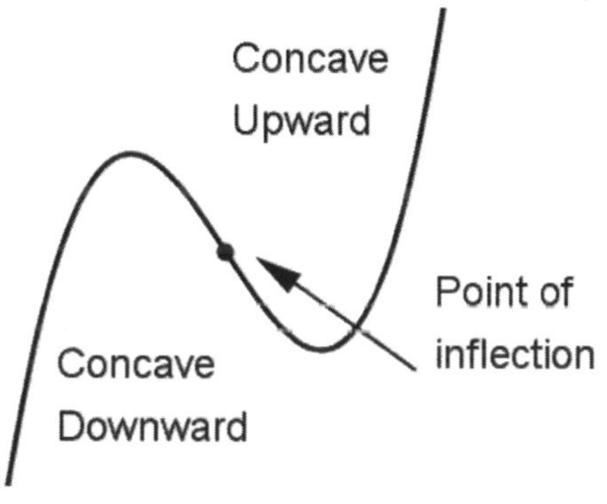

It is important to note in situations where $f'(p) = f''(p) = 0$ it is necessary to use the first derivative test to determine if p is an inflection point.

The derivative tests that have been discussed thus far can help you get a rough picture of what the graph of an unfamiliar function looks like. Begin by solving the equation $f(x) = 0$ to find all the zeros of the function, if they exist. Plot these points on the graph. Then, find the first derivative of the function and solve the equation $f'(x) = 0$ to find the critical points. Remember the numbers obtained here are the x portions of the coordinates. Substitute these values for x in the original function and solve for y to get the full coordinates of the points. Plot these points on the graph. Take the second derivative of the function and solve the equation $f''(x) = 0$ to find the points of inflection. Substitute in the original function to get the coordinates and graph these points. Test points on both sides of the critical points to test for concavity and draw the curve.

Review Video: Second Derivative Test
Visit mometrix.com/academy and enter code: 655614

Derivative Problems

A derivative represents the rate of change of a function; thus, derivatives are a useful tool for solving any problem that involves finding the rate at which a function is changing. In its simplest form, such a problem might provide a formula for a quantity as a function of time and ask for its rate of change at a particular time.

If the temperature in a chamber in degrees Celsius is equal to $T(t) = 20 + e^{-\left(\frac{t}{2}\right)}$, where t is the time in seconds, then the derivative of the function represents the rate of change of the temperature over time. The rate of change is equal to $\frac{dT}{dt} = \frac{d}{dt}\left(20 + e^{-\left(\frac{t}{2}\right)}\right) = -\frac{1}{2}e^{-\left(\frac{t}{2}\right)}$, and the initial rate of change is $T'(0) = -\frac{1}{2}e^{-\left(\frac{0}{2}\right)} = -\frac{1}{2}\frac{°C}{s}$.

Suppose we are told that the net profit that a small company makes when it produces and sells x units of a product is equal to $P(x) = 200x - 20{,}000$. The derivative of this function would be the *additional profit for each additional unit sold*, a quantity known as the marginal profit. The marginal profit in this case is $P'(x) = 200$.

Solving Related Rates Problems

A **related rate problem** is one in which one variable has a relation with another variable, and the rate of change of one of the variables is known. With that information, the rate of change of the other variable can be determined. The first step in solving related rates problems is defining the known rate of change. Then, determine the relationship between the two variables, then the derivatives (the rates of change), and finally substitute the problem's specific values. Consider the following example:

The side of a cube is increasing at a rate of 2 feet per second. Determine the rate at which the volume of the cube is increasing when the side of the cube is 4 feet long.

For the problem in question, the known rate of change can be expressed as $s'(t) = \frac{ds}{dt} = 2\frac{\text{ft}}{\text{s}}$, where s is the length of the side and t is the elapsed time in seconds. The relationship between the two variables of the cube is $v = s^3$, where v is the volume of the cube and s is the length of the side. The unknown rate of change to determine is the volume. As both v and s change with time, $v = s^3$ becomes $v(t) = [s(t)]^3$

Now, the chain rule is applied to differentiate both sides of the equation with respect to t.

$$d\frac{v(t)}{dt} = \frac{d[s(t)]^3}{dt};\ \frac{dv}{dt} = \frac{(3[s(t)]^2)ds}{dt}$$

Finally, the specific value of $s = 4$ feet is substituted, and the equation is evaluated.

$$\begin{aligned}\frac{dv}{dt} &= \frac{(3[s(t)]^2)ds}{dt} \\ &= 3(4)^2 \times 2 = 96\frac{\text{ft}^3}{\text{s}} \\ &= 96 \text{ cubic}\frac{\text{ft}}{\text{s}}\end{aligned}$$

Therefore, when a side of the cube is 4 feet long, the volume of the cube is increasing at a rate of 96 cubic $\frac{\text{feet}}{\text{second}}$.

Review Video: Solving Related Rates Problems
Visit mometrix.com/academy and enter code: 321959

Solving Optimization Problems

An **optimization problem** is a problem in which we are asked to find the value of a variable that maximizes or minimizes a particular value. Because the minimum or maximum occurs at a critical point, and because the critical point occurs when the derivative of the function is zero, we can solve an optimization problem by setting the derivative of the function to zero and solving for the desired variable.

For example, suppose a farmer has 720 m of fencing, and wants to use it to fence in a 2 by 3 block of identical rectangular pens. What dimensions of the pens will maximize their area?

We can draw a diagram:

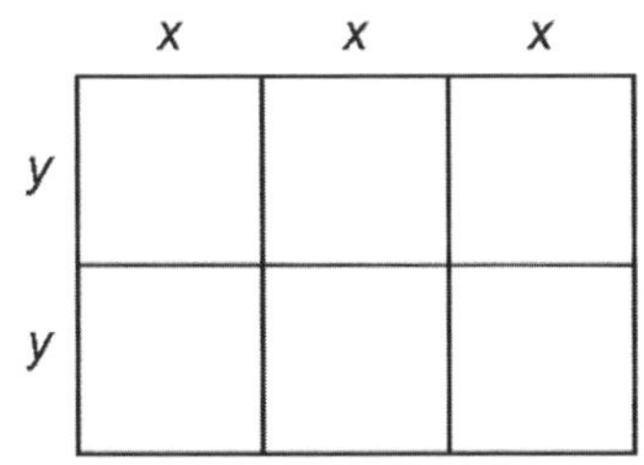

We want to maximize the area of each pen, $A(x, y) = xy$. However, we have the additional constraint that the farmer has only 720 m of fencing. In terms of x and y, we can count the number of segments of each length, 9 for x and 8 for y, so the total amount of fencing required will be $9x + 8y$. Our constraint becomes $9x + 8y = 720$; solving for y yields $y = -\frac{9}{8}x + 90$. We can substitute that into the area equation to get $A(x) = x\left(-\frac{9}{8}x + 90\right) = -\frac{9}{8}x^2 + 90x$. Taking the derivative yields $A'(x) = -\frac{9}{4}x + 90$; setting that equal to zero and solving for x yields $x = 40$. $y = -\frac{9}{8}(40) + 90 = 45$; thus, the maximum dimensions of the pen are 40 by 45 meters.

Review Video: Solving Optimization Problems
Visit mometrix.com/academy and enter code: 628734

Characteristics of Functions (Using Calculus)

Rolle's theorem states that if a differentiable function has two different values in the domain that correspond to a single value in the range, then the function must have a point between them where the slope of the tangent to the graph is zero. This point will be a maximum or a minimum value of the function between those two points. The maximum or minimum point is the point at which $f'(c) = 0$, where c is within the appropriate interval of the function's domain. The following graph shows a function with one maximum in the second quadrant and one minimum in the fourth quadrant.

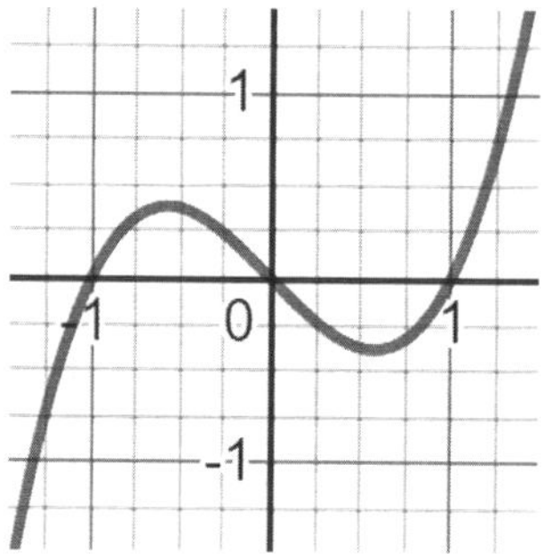

MEAN VALUE THEOREM

According to the **mean value theorem**, between any two points on a curve, there exists a tangent to the curve whose slope is parallel to the chord formed by joining those two points. Remember the formula for slope: $m = \frac{\Delta y}{\Delta x}$. In a function, $f(x)$ represents the value for y. Therefore, if you have two points on a curve, m and n, the corresponding points are $(m, f(m))$ and $(n, f(n))$. Assuming $m < n$, the formula for the slope of the chord joining those two points is $\frac{f(n)-f(m)}{n-m}$. This must also be the slope of a line parallel to the chord, since parallel lines have equal slopes. Therefore, there must be a value p between m and n such that $f'(p) = \frac{f(n)-f(m)}{n-m}$.

For a function to have continuity, its graph must be an unbroken curve. That is, it is a function that can be graphed without having to lift the pencil to move it to a different point. To say a function is continuous at point p, you must show the function satisfies three requirements. First, $f(p)$ must exist. If you evaluate the function at p, it must yield a real number. Second, there must exist a relationship such that $\lim_{x \to p} f(x) = f(p)$. Finally, the following relationship must be true:

$$\lim_{x \to p^+} F(x) = \lim_{x \to p^-} F(x) = F(p)$$

If all three of these requirements are met, a function is considered continuous at p. If any one of them is not true, the function is not continuous at p.

Review Video: Mean Value Theorem
Visit mometrix.com/academy and enter code: 633482

TANGENTS

Tangents are lines that touch a curve in exactly one point and have the same slope as the curve at that point. To find the slope of a curve at a given point and the slope of its tangent line at that point, find the derivative of the function of the curve. If the slope is undefined, the tangent is a vertical line. If the slope is zero, the tangent is a horizontal line.

A line that is normal to a curve at a given point is perpendicular to the tangent at that point. Assuming $f'(x) \neq 0$, the equation for the normal line at point (a, b) is: $y - b = -\frac{1}{f'(a)}(x - a)$. The easiest way to find the slope of the normal is to take the negative reciprocal of the slope of the tangent. If the slope of the tangent is zero, the slope of the normal is undefined. If the slope of the tangent is undefined, the slope of the normal is zero.

ANTIDERIVATIVES (INTEGRALS)

The antiderivative of a function is the function whose first derivative is the original function. Antiderivatives are typically represented by capital letters, while their first derivatives are represented by lower case letters. For example, if $F' = f$, then F is the antiderivative of f. Antiderivatives are also known as indefinite integrals. When taking the derivative of a function, any constant terms in the function are eliminated because their derivative is 0. To account for this possibility, when you take the indefinite integral of a function, you must add an unknown constant C to the end of the function. Because there is no way to know what the value of the original constant was when looking just at the first derivative, the integral is indefinite.

To find the indefinite integral, reverse the process of differentiation. Below are the formulas for constants and powers of x.

$$\int 0\,dx = C$$

$$\int k\,dx = kx + C$$

$$\int x^n\,dx = \frac{x^{n+1}}{n+1} + C, \text{where } n \neq -1$$

Recall that in the differentiation of powers of x, you multiplied the coefficient of the term by the exponent of the variable and then reduced the exponent by one. In integration, the process is reversed: add one to the value of the exponent, and then divide the coefficient of the term by this number to get the integral. Because you do not know the value of any constant term that might have been in the original function, add C to the end of the function once you have completed this process for each term.

Review Video: Indefinite Integrals
Visit mometrix.com/academy and enter code: 541913

Finding the integral of a function is the opposite of finding the derivative of the function. Where possible, you can use the trigonometric or logarithmic differentiation formulas in reverse, and add C to the end to compensate for the unknown term. In instances where a negative sign appears in the differentiation formula, move the negative sign to the opposite side (multiply both sides by -1) to reverse for the integration formula. You should end up with the following formulas:

$$\int \cos x\ dx = \sin x + C$$

$$\int \sec x \tan x\ dx = \sec x + C$$

$$\int \sin x\ dx = -\cos x + C$$

$$\int \csc x \cot x\ dx = -\csc x + C$$

$$\int \sec^2 x\ dx = \tan x + C$$

$$\int \csc^2 x\ dx = -\cot x + C$$

$$\int \frac{1}{x}\ dx = \ln|x| + C$$

$$\int e^x\ dx = e^x + C$$

Integration by substitution is the integration version of the chain rule for differentiation. The formula for integration by substitution is given by the equation

$$\int f(g(x))g'(x)dx = \int f(u)du\,;\quad u = g(x) \text{ and } du = g'(x)dx$$

When a function is in a format that is difficult or impossible to integrate using traditional integration methods and formulas due to multiple functions being combined, use the formula shown above to convert the function to a simpler format that can be integrated directly.

Integration by parts is the integration version of the product rule for differentiation. Whenever you are asked to find the integral of the product of two different functions or parts, integration by parts can make the process simpler. Recall for differentiation $(fg)'(x) = f(x)g'(x) + g(x)f'(x)$. This can also be written $\frac{d}{dx}(u \times v) = u\frac{dv}{dx} + v\frac{du}{dx}$, where $u = f(x)$ and $v = g(x)$. Rearranging to integral form gives the formula:

$$\int u\,dv = uv - \int v\,du$$

$$\int f(x)g'(x)\,dx = f(x)g(x) - \int f'(x)g(x)\,dx$$

When using integration by parts, the key is selecting the best functions to substitute for u and v so that you make the integral easier to solve and not harder.

While the indefinite integral has an undefined constant added at the end, the definite integral can be calculated as an exact real number. To find the definite integral of a function over a closed interval, use the formula $\int_n^m f(x)\,dx = F(m) - F(n)$ where F is the integral of f. Because you have been given the boundaries of n and m, no undefined constant C is needed.

Review Video: Integration by Parts
Visit mometrix.com/academy and enter code: 459972

Review Video: Integration by Substitution
Visit mometrix.com/academy and enter code: 740649

First Fundamental Theorem of Calculus

The **first fundamental theorem of calculus** shows that the process of indefinite integration can be reversed by finding the first derivative of the resulting function. It also gives the relationship between differentiation and integration over a closed interval of the function. For example, assuming a function is continuous over the interval $[m, n]$, you can find the definite integral by using the formula

$$\int_m^n f(x)\,dx = F(n) - F(m)$$

To find the **average value** of the function over the given interval, use the formula:

$$\frac{1}{n-m}\int_m^n f(x)\,dx$$

Review Video: First Fundamental Theorem of Calculus
Visit mometrix.com/academy and enter code: 248431

Second Fundamental Theorem of Calculus

The **second fundamental theorem of calculus** is related to the first. This theorem states that, assuming the function is continuous over the interval you are considering, taking the derivative of the integral of a function will yield the original function. The general format for this theorem for any point having a domain value equal to c in the given interval is:

$$\frac{d}{dx}\int_c^x f(t)\,dt = f(x)$$

For each of the following **properties of integrals** of function f, the variables m, n, and p represent values in the domain of the given interval of $f(x)$. The function is assumed to be integrable across all relevant intervals.

Swapping the limits of integration:	$\int_m^n f(x)\,dx = -\int_n^m f(x)\,dx$
Function multiplied by a constant:	$\int_m^n kf(x)dx = k\int_m^n f(x)\,dx$
Separating the integral into parts:	$\int_m^n f(x)\,dx = \int_m^p f(x)\,dx + \int_p^n f(x)\,dx$
If the limits of integration are equivalent:	$\int_n^n f(x)\,dx = 0$
If $f(x)$ is an even function and the limits of integration are symmetric:	$\int_{-m}^m f(x)\,dx = 2\int_0^m f(x)\,dx$
If $f(x)$ is an odd function and the limits of integration are symmetric:	$\int_{-m}^m f(x)\,dx = 0$

Review Video: Second Fundamental Theorem of Calculus
Visit mometrix.com/academy and enter code: 524689

Matching Functions to Derivatives or Accumulations

Derivatives

We can use what we know about the meaning of a derivative to match the graph of a function with a graph of its derivative. For one thing, we know that where the function has a critical point, the derivative is zero. Therefore, at every x value at which the graph of a function has a maximum or minimum, the derivative must cross the *x-axis*—and conversely, everywhere the graph of the derivative crosses the x axis, the function must have a critical point: either a maximum, a minimum, or an inflection point. If this is still not enough to identify the correct match, we can also use the fact that the sign of the derivative corresponds to whether the function is increasing or decreasing:

everywhere the graph of the derivative is above the x-axis, the function must be increasing (its slope is positive), and everywhere the graph of the derivative is below the x-axis, the function must be decreasing (its slope is negative).

For example, below are graphs of the function and its derivative. The maxima and minima of the function (left) are circled, and the zeroes of the derivative (right) are circled.

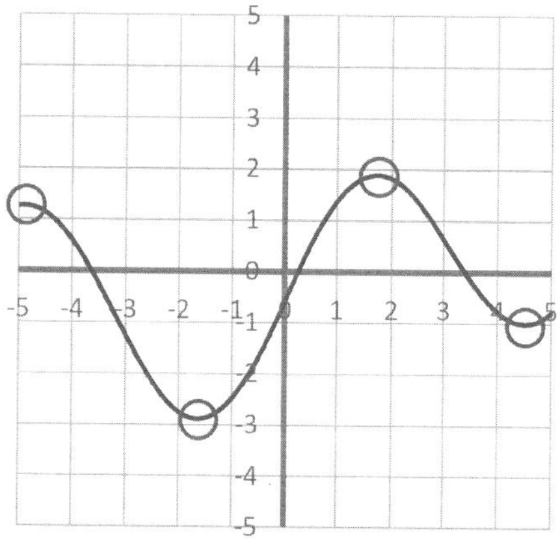

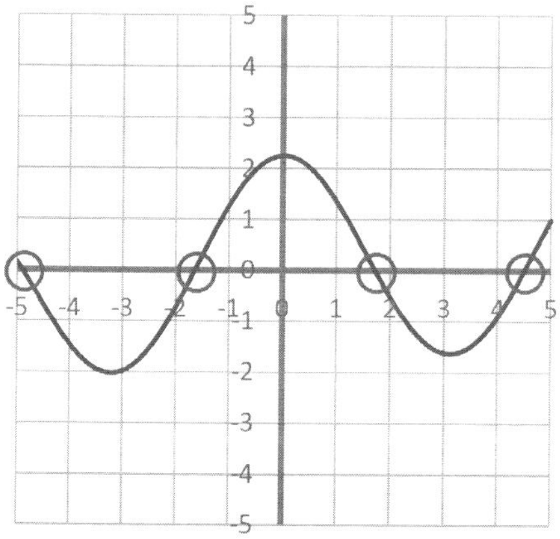

Accumulations

The **accumulation** of a function is another name for its antiderivative, or integral. We can use the relationship between a function and its antiderivative to match the corresponding graphs. For example, we know that where the graph of the function is above the x-axis, the function is positive, thus the accumulation must be increasing (its slope is positive); where the graph of the function is below the x-axis, the accumulation must be decreasing (its slope is negative). It follows that where the function changes from positive to negative—where the graph crosses the x-axis with a negative slope—, its accumulation changes from increasing to decreasing—so the accumulation has a local maximum. Where the function changes from negative to positive—where its graph crosses the x-axis with a positive slope—, the accumulation has a local minimum.

For example, below are graphs of a function and its accumulation. The points on the function (left) where the graph crosses the x-axis are circled; the local minima and maxima of the accumulation (right) are circled.

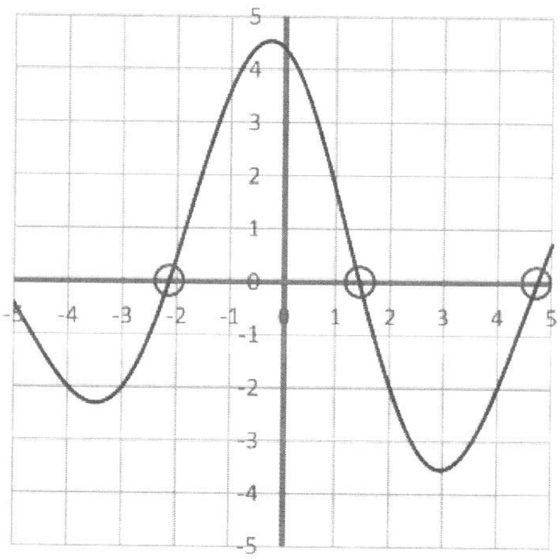

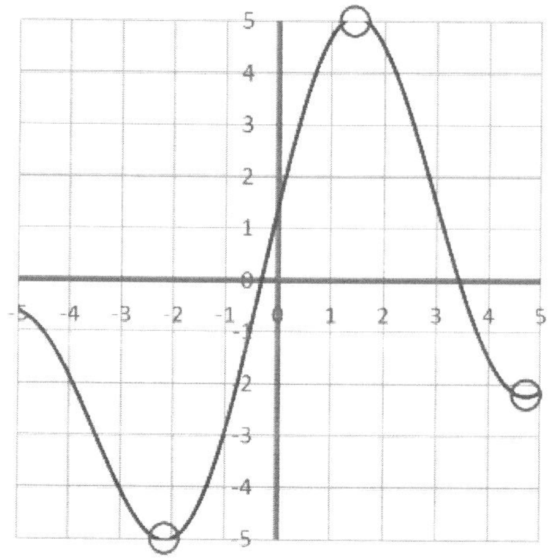

RIEMANN SUMS

A **Riemann sum** is a sum used to approximate the definite integral of a function over a particular interval by dividing the area under the function into vertical rectangular strips and adding the areas of the strips. The height of each strip is equal to the value of the function at some point within the interval covered by the strip. Formally, if we divide the interval over which we are finding the area into n intervals bounded by the $n+1$ points $\{x_i\}$ (where x_0 and x_n are the left and right bounds of the interval), then the Riemann sum is $\sum_{i=1}^{n} f(x_i^*)\Delta x_i$, where $\Delta x_i = x_i - x_{i-1}$ and x_i^* is some point in the interval $[x_{i-1}, x_i]$. In principle, any point in the interval can be chosen, but common choices include the left endpoint of the interval (yielding the **left Riemann sum**), the right endpoint (yielding the **right Riemann sum**), and the midpoint of the interval (the basis of the **midpoint rule**). Usually, it is convenient to set all the intervals to the same width, although the definition of the Riemann sum does not require this.

The following graphic shows the rectangular strips used for one possible Riemann sum of a particular function:

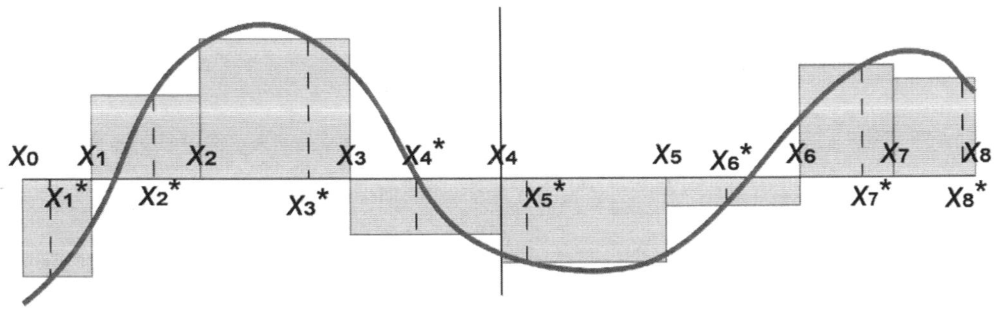

LEFT AND RIGHT RIEMANN SUMS

A **Riemann sum** is an approximation to the definite integral of a function over a particular interval performed by dividing it into smaller intervals and summing the products of the width of each interval and the value of the function evaluated at some point within the interval. The **left Riemann sum** is a Riemann sum in which the function is evaluated at the left endpoint of each interval. In the **right Riemann sum**, the function is evaluated at the right endpoint of each interval.

Review Video: Left and Right Riemann Sums
Visit mometrix.com/academy and enter code: 998275

When the function is increasing, the left Riemann sum will always underestimate the function. This is because we are evaluating the function at the minimum point within each interval; the integral of the function in the interval will be larger than the estimate. Conversely, the right Riemann sum is evaluating the function at the maximum point within each interval, thus it will always overestimate the function. Consider the following diagrams, in which the area under the same increasing function is shown approximated by a left Riemann sum and a right Riemann sum:

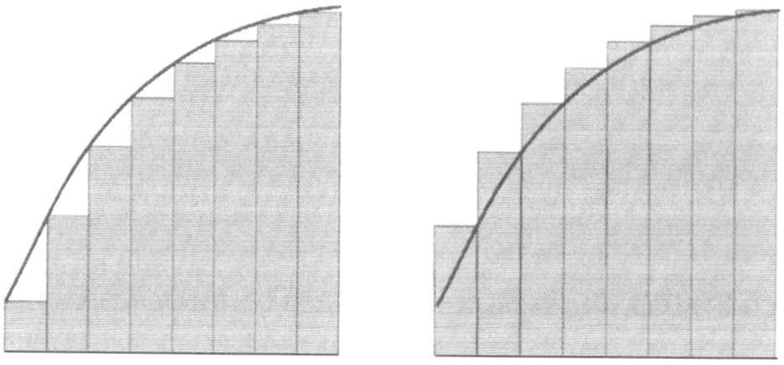

For a decreasing function these considerations are reversed: a left Riemann sum will overestimate the integral, and a right Riemann sum will underestimate it.

Midpoint Rule

The **midpoint rule** is a way of approximating the definite integral of a function over an interval by dividing the interval into smaller sub-intervals, multiplying the width of each sub-interval by the value of the function at the midpoint of the sub-interval, and then summing these products. This is a special case of the Riemann sum, specifying the midpoint of the interval as the point at which the function is to be evaluated. The approximation found using the midpoint rule is usually more accurate than that found using the left or right Riemann sum, though as the number of intervals becomes very large the difference becomes negligible.

For example, suppose we are asked to estimate by the midpoint rule the integral of $f(x) = \frac{1}{x}$ in the interval [2, 4]. We can divide this interval into four intervals of width $\frac{1}{2}$: [2, 2.5], [2.5, 3], [3, 3.5], and [3.5, 4]. (The more intervals, the more accurate the estimate, but we'll use a small number of intervals in this example to keep it simple.) The midpoint rule then gives an estimate of $\frac{1}{2}\big(f(2.25)\big) + \frac{1}{2}\big(f(2.75)\big) + \frac{1}{2}\big(f(3.25)\big) + \frac{1}{2}\big(f(3.75)\big) = \frac{1}{2}\left(\frac{4}{9}\right) + \frac{1}{2}\left(\frac{4}{11}\right) + \frac{1}{2}\left(\frac{4}{13}\right) + \frac{1}{2}\left(\frac{4}{15}\right) \approx 0.691$, not far from the actual value of $\int_2^4 \frac{1}{x}dx = [\ln x]_2^4 \approx 0.693$.

Review Video: Midpoint Rule
Visit mometrix.com/academy and enter code: 790070

Trapezoid Rule

The **trapezoid rule** is a method of approximating the definite integral of a function by dividing the area under the function into a series of trapezoidal strips, the upper corners of the trapezoid touching the function, and adding the areas of the strips. The following diagram shows the use of the trapezoid rule to estimate the integral of the function $y = 2^x$ in the interval $[0, 3]$:

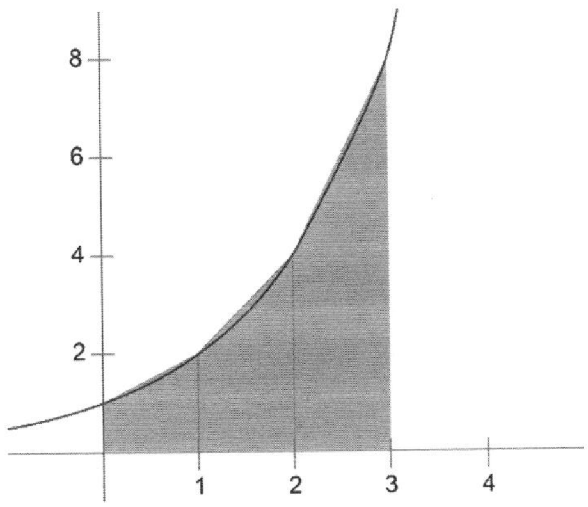

Mathematically, if we define the endpoints of the n subdivisions to be $\{x_{i-1}, x_i\}$: $(1 \leq i \leq n)$, where x_0 and x_n are the endpoints of the entire interval over which we are estimating the integral, then the result of the application of the trapezoid rule is equal to $\sum_{i=1}^{n}\left(\frac{f(x_{i-1})+f(x_i)}{2}\right)(x_i - x_{i-1})$. For the example shown above, that yields $\left(\frac{2^1+2^0}{2}\right)(1-0) + \left(\frac{2^2+2^1}{2}\right)(2-1) + \left(\frac{2^3+2^2}{2}\right)(3-2) = \frac{21}{2}$, or 10.5—not far from the actual value of $\int_0^3 2^x dx = \int_0^3 e^{x \ln 2}\, dx = \left[\frac{2^x}{\ln 2}\right]_0^3 \approx 10.1$. (Of course, we could have achieved more accuracy by using smaller subdivisions.)

The trapezoid rule is related to the Riemann sum, but usually gives more accurate results than the left or right Riemann sum for the same number of intervals. In fact, it isn't hard to prove that the answer given by the trapezoid rule is equal to the average of the left and right Riemann sums using the same partition.

Review Video: Trapezoid Rule
Visit mometrix.com/academy and enter code: 170482

LIMIT OF RIEMANN SUMS

As the number of sub-intervals becomes larger, and the width of each sub-interval becomes smaller, the approximation becomes increasingly accurate, and at the limit as the number of sub-intervals approaches infinity and their width approaches zero, the value becomes exact. In fact, the definite integral is often defined as a limit of Riemann sums.

It's possible to find the definite integral by this method. Suppose we want to find the integral of $f(x) = x^2$ over the interval [0, 2]. We'll divide this interval into n sub-intervals of equal width and evaluate the function at the right endpoint of each sub-interval. (This choice is arbitrary; at the limit the answer would be the same if we chose the left endpoint, or any other point within the interval.) Our Riemann sum becomes $\sum_{i=1}^{n}\frac{2}{n}\left(\frac{2}{n}i\right)^2 = \frac{8}{n^3}\sum_{i=1}^{n} i^2$, where $\sum_{i=1}^{n} i^2 = \frac{1}{6}n(n+1)(2n+1)$, thus this becomes $\frac{8}{n^3}\cdot\frac{1}{6}n(n+1)(2n+1) = \frac{4}{3}\left(1+\frac{1}{n}\right)\left(2+\frac{1}{n}\right)$. At the limit as $n \to \infty$, this becomes $\frac{4}{3}(1)(2) = \frac{8}{3}$. This is the same result as we get by integrating directly: $\int_0^2 x^2 dx = \left[\frac{1}{3}x^3\right]_0^2 = \frac{1}{3}(2)^3 - \frac{1}{3}(0)^3 = \frac{8}{3}$.

Uses for Integration

Calculating Distances

When given the velocity of an object over time, it's possible to find a distance by integration. The velocity is the rate of change of the position; therefore, the displacement is the accumulation of the velocity: that is, the integral of the velocity is the displacement. However, if asked to find the total distance traveled (as opposed to the displacement), it's important to take the sign into account: we must integrate not just the velocity, but the absolute value of the velocity, which essentially means integrating separately over each interval in which the velocity has a different sign.

For example, suppose we're asked to find the total distance traveled from $t = 0$ to $t = 8$ by an object moving with a velocity in meters per second given by the equation $v(t) = 2\sqrt{t} - t$. This function is zero when $2\sqrt{t} - t = 0 \Rightarrow \sqrt{t}(2 - \sqrt{t}) = 0 \Rightarrow t = 0$ or 4. $v(t)$ is positive when $0 < t < 4$ and negative when $t > 4$. Thus, the distance travelled is

$$\begin{aligned}
\int_0^8 |v(t)|dt &= \int_0^8 \left|2\sqrt{t} - t\right|dt \\
&= \int_0^4 \left(2\sqrt{t} - t\right)dt - \int_4^8 \left(2\sqrt{t} - t\right)dt \\
&= \left[\frac{4}{3}t^{\frac{3}{2}} - \frac{1}{2}t^2\right]_0^4 - \left[\frac{4}{3}t^{\frac{3}{2}} - \frac{1}{2}t^2\right]_4^8 \\
&= \left(\left[\frac{4}{3}(4)^{\frac{3}{2}} - \frac{1}{2}4^2\right] - \left[\frac{4}{3}(0)^{\frac{3}{2}} - \frac{1}{2}(0)^2\right]\right) - \left(\left[\frac{4}{3}(8)^{\frac{3}{2}} - \frac{1}{2}(8)^2\right] - \left[\frac{4}{3}(4)^{\frac{3}{2}} - \frac{1}{2}(4)^2\right]\right) \\
&= \left(\left[\frac{32}{3} - 8\right] - 0\right) - \left(\left[\frac{64}{3}\sqrt{2} - 32\right] - \left[\frac{32}{3} - 8\right]\right) \\
&= \frac{8}{3} - \frac{64}{3}\sqrt{2} + \frac{96}{3} + \frac{8}{3} \\
&= \frac{8}{3}\left(1 - 8\sqrt{2} + 12 + 1\right) \\
&= \frac{8}{3}\left(14 - 8\sqrt{2}\right) \approx 7.16 \text{ meters}
\end{aligned}$$

Review Video: Calculating Distances Using Integration
Visit mometrix.com/academy and enter code: 866719

Calculating Areas

One way to calculate the area of an irregular shape is to find a formula for the width of the shape along the x direction as a function of the y coordinate, and then integrate over y, or vice versa. What this amounts to is dividing the area into thin strips and adding the areas of the strips—and then taking the limit as the width of the strips approaches zero.

For example, suppose we want to find the area enclosed by the functions $y_1 = x^2$ and $y_2 = (2 - x^2)$. The height of this enclosure is equal to $y_2 - y_1 = 2 - 2x^2$; we can find the area by integrating this height over x. The two shapes intersect at the points $(1, 1)$ and $(-1, 1)$, thus our limits of integration are -1 and 1. Thus the area can be found as:

$$\int_{-1}^{1}(2-2x^2)dx=\left[2x-\frac{2}{3}x^3\right]_{-1}^{1}=\left(2(1)-\frac{2}{3}(1)\right)-\left(2(-1)-\frac{2}{3}(-1)\right)=\frac{8}{3}$$

Review Video: Calculating Areas Using Integration
Visit mometrix.com/academy and enter code: 118949

Calculating Volumes

One way to calculate the volume of a three-dimensional shape is to find a formula for its cross-sectional area perpendicular to some axis and then integrate over that axis. Effectively, this divides the shape into thin, flat slices and adds the volumes of the slices—and then takes the limit as the thickness of the slices approaches zero.

For example, suppose we want to find the volume of the ellipsoid $4x^2+4y^2+z^2=36$. If we take a cross-section perpendicular to the z-axis, this has the formula $4x^2+4y^2=36-z^2$, or $x^2+y^2=9-\frac{z^2}{4}$; this is the formula of a circle with a radius of $\sqrt{9-\frac{z^2}{4}}$, and thus has an area of $\pi\left(9-\frac{z^2}{4}\right)$. To find the volume, we integrate this formula over z. The maximum and minimum values of z occur when $x=y=0$, and then $z^2=36$, thus $z=\pm 6$; these are our limits of integration. Thus, the volume is:

$$\begin{aligned}\int_{-6}^{6}\pi\left(9-\frac{z^2}{4}\right)dz&=\pi\left[9z-\frac{z^3}{12}\right]_{-6}^{6}\\&=\pi\left[\left(9(6)-\frac{6^3}{12}\right)-\left(9(-6)-\frac{(-6)^3}{12}\right)\right]\\&=\pi[(54-18)-(-54+18)]\\&=72\pi\approx 226.2\end{aligned}$$

Review Video: Calculating Volume Using Integration
Visit mometrix.com/academy and enter code: 100341

Chapter Quiz

Ready to see how well you retained what you just read? Scan the QR code to go directly to the chapter quiz interface for this study guide. If you're using a computer, simply visit the online resources page at **mometrix.com/resources719/orelamath** and click the Chapter Quizzes link.

Statistics, Probability, and Discrete Mathematics

Displaying Data

Displaying Information

Frequency Tables

Frequency tables show how frequently each unique value appears in a set. A **relative frequency table** is one that shows the proportions of each unique value compared to the entire set. Relative frequencies are given as percentages; however, the total percent for a relative frequency table will not necessarily equal 100 percent due to rounding. An example of a frequency table with relative frequencies is below.

Favorite Color	Frequency	Relative Frequency
Blue	4	13%
Red	7	22%
Green	3	9%
Purple	6	19%
Cyan	12	38%

Review Video: Data Interpretation of Graphs
Visit mometrix.com/academy and enter code: 200439

Circle Graphs

Circle graphs, also known as *pie charts*, provide a visual depiction of the relationship of each type of data compared to the whole set of data. The circle graph is divided into sections by drawing radii to create central angles whose percentage of the circle is equal to the individual data's percentage of the whole set. Each 1% of data is equal to 3.6° in the circle graph. Therefore, data represented by a 90° section of the circle graph makes up 25% of the whole. When complete, a circle graph often looks like a pie cut into uneven wedges. The pie chart below shows the data from the frequency table referenced earlier where people were asked their favorite color.

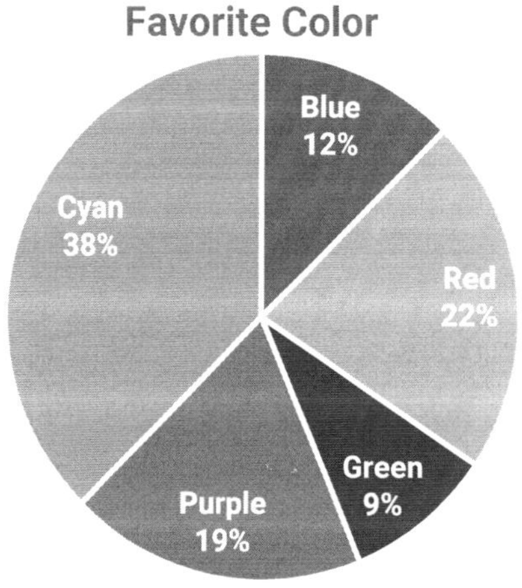

PICTOGRAPHS

A **pictograph** is a graph, generally in the horizontal orientation, that uses pictures or symbols to represent the data. Each pictograph must have a key that defines the picture or symbol and gives the quantity each picture or symbol represents. Pictures or symbols on a pictograph are not always shown as whole elements. In this case, the fraction of the picture or symbol shown represents the same fraction of the quantity a whole picture or symbol stands for. For example, a row with $3\frac{1}{2}$ ears of corn, where each ear of corn represents 100 stalks of corn in a field, would equal $3\frac{1}{2} \times 100 = 350$ stalks of corn in the field.

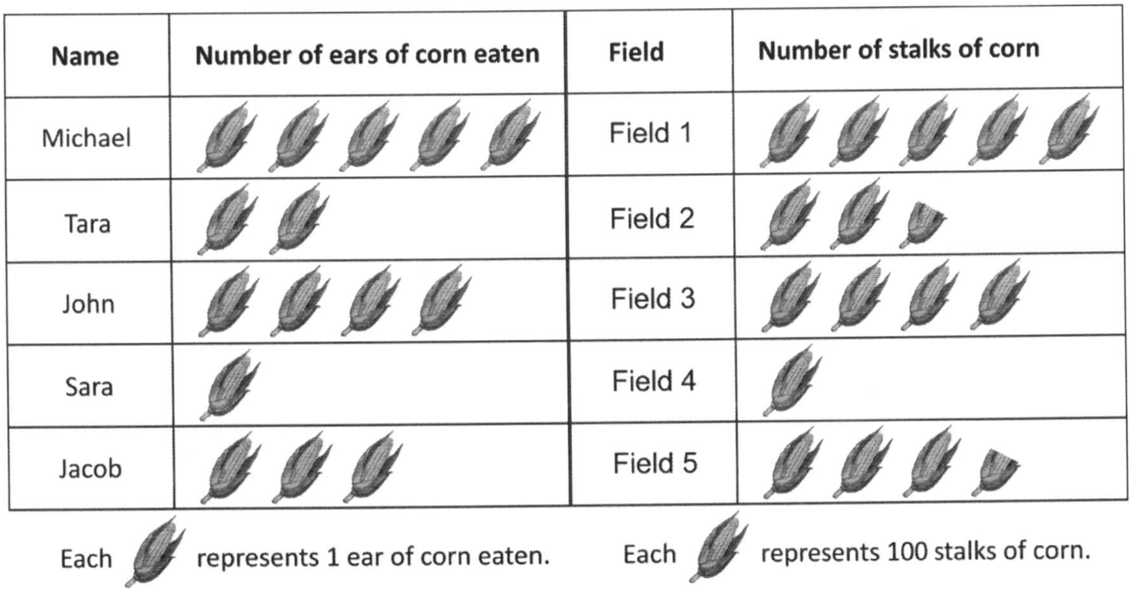

Review Video: Pictographs
Visit mometrix.com/academy and enter code: 147860

LINE GRAPHS

Line graphs have one or more lines of varying styles (solid or broken) to show the different values for a set of data. The individual data are represented as ordered pairs, much like on a Cartesian plane. In this case, the x- and y-axes are defined in terms of their units, such as dollars or time. The individual plotted points are joined by line segments to show whether the value of the data is increasing (line sloping upward), decreasing (line sloping downward), or staying the same (horizontal line). Multiple sets of data can be graphed on the same line graph to give an easy visual comparison. An example of this would be graphing achievement test scores for different groups of students over the same time period to see which group had the greatest increase or decrease in performance from year to year (as shown below).

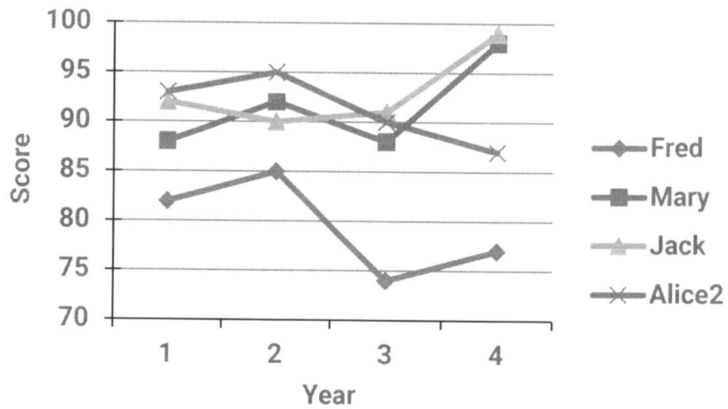

Review Video: How to Create a Line Graph
Visit mometrix.com/academy and enter code: 480147

Line Plots

A **line plot**, also known as a *dot plot*, has plotted points that are not connected by line segments. In this graph, the horizontal axis lists the different possible values for the data, and the vertical axis lists the number of times the individual value occurs. A single dot is graphed for each value to show the number of times it occurs. This graph is more closely related to a bar graph than a line graph. Do not connect the dots in a line plot or it will misrepresent the data.

Review Video: Line Plot
Visit mometrix.com/academy and enter code: 754610

Stem and Leaf Plots

A **stem and leaf plot** is useful for depicting groups of data that fall into a range of values. Each piece of data is separated into two parts: the first, or left, part is called the stem; the second, or right, part is called the leaf. Each stem is listed in a column from smallest to largest. Each leaf that has the common stem is listed in that stem's row from smallest to largest. For example, in a set of two-digit numbers, the digit in the tens place is the stem, and the digit in the ones place is the leaf. With a stem and leaf plot, you can easily see which subset of numbers (10s, 20s, 30s, etc.) is the largest. This information is also readily available by looking at a histogram, but a stem and leaf plot also allows you to look closer and see exactly which values fall in that range. Using a sample set of test scores (82, 88, 92, 93, 85, 90, 92, 95, 74, 88, 90, 91, 78, 87, 98, 99), we can assemble a stem and leaf plot like the one below.

Test Scores

7	4	8							
8	2	5	7	8	8				
9	0	0	1	2	2	3	5	8	9

Review Video: Stem and Leaf Plots
Visit mometrix.com/academy and enter code: 302339

Bar Graphs

A **bar graph** is one of the few graphs that can be drawn correctly in two different configurations – both horizontally and vertically. A bar graph is similar to a line plot in the way the data is organized on the graph. Both axes must have their categories defined for the graph to be useful. Rather than placing a single dot to mark the point of the data's value, a bar, or thick line, is drawn from zero to the exact value of the data, whether it is a number, percentage, or other numerical value. Longer bar lengths correspond to greater data values. To read a bar graph, read the labels for the axes to find the units being reported. Then, look where the bars end in relation to the scale given on the corresponding axis and determine the associated value.

The bar chart below represents the responses from our favorite-color survey.

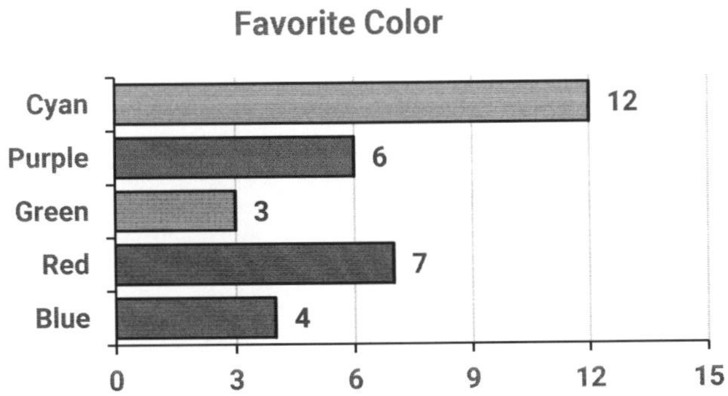

Histograms

At first glance, a **histogram** looks like a vertical bar graph. The difference is that a bar graph has a separate bar for each piece of data and a histogram has one continuous bar for each *range* of data. For example, a histogram may have one bar for the range 0–9, one bar for 10–19, etc. While a bar graph has numerical values on one axis, a histogram has numerical values on both axes. Each range is of equal size, and they are ordered left to right from lowest to highest. The height of each column on a histogram represents the number of data values within that range. Like a stem and leaf plot, a histogram makes it easy to glance at the graph and quickly determine which range has the greatest quantity of values. A simple example of a histogram is below.

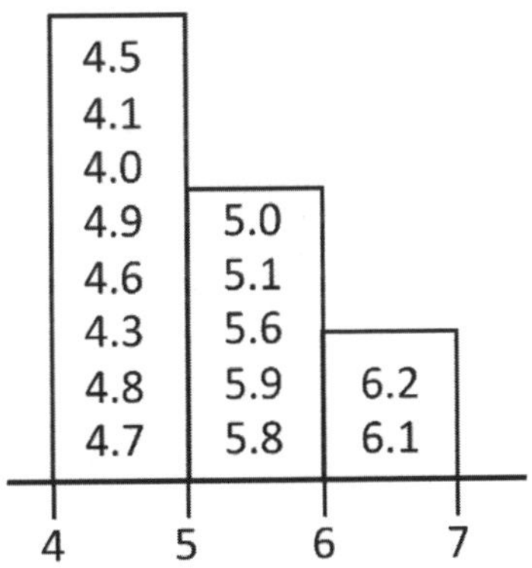

5-Number Summary

The **5-number summary** of a set of data gives a very informative picture of the set. The five numbers in the summary include the minimum value, maximum value, and the three quartiles. This information gives the reader the range and median of the set, as well as an indication of how the data is spread about the median.

BOX AND WHISKER PLOTS

A **box-and-whiskers plot** is a graphical representation of the 5-number summary. To draw a box-and-whiskers plot, plot the points of the 5-number summary on a number line. Draw a box whose ends are through the points for the first and third quartiles. Draw a vertical line in the box through the median to divide the box in half. Draw a line segment from the first quartile point to the minimum value, and from the third quartile point to the maximum value.

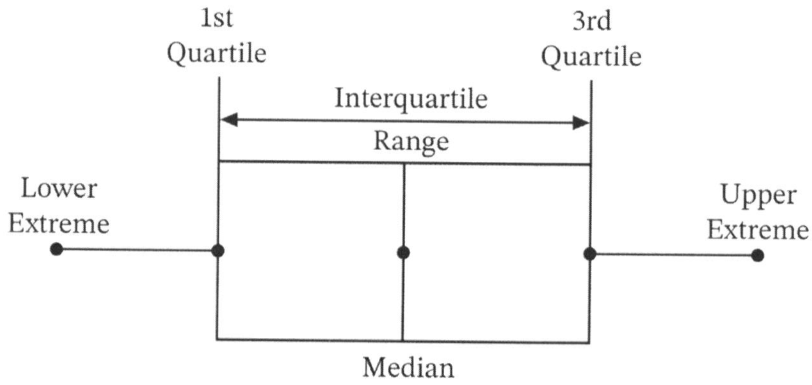

Review Video: Box and Whisker Plots
Visit mometrix.com/academy and enter code: 810817

EXAMPLE

Given the following data (32, 28, 29, 26, 35, 27, 30, 31, 27, 32), we first sort it into numerical order: 26, 27, 27, 28, 29, 30, 31, 32, 32, 35. We can then find the median. Since there are ten values, we take the average of the 5th and 6th values to get 29.5. We find the lower quartile by taking the median of the data smaller than the median. Since there are five values, we take the 3rd value, which is 27. We find the upper quartile by taking the median of the data larger than the overall median, which is 32. Finally, we note our minimum and maximum, which are simply the smallest and largest values in the set: 26 and 35, respectively. Now we can create our box plot:

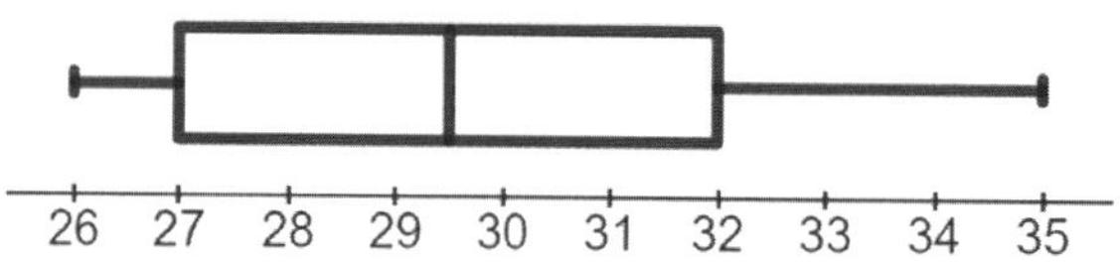

This plot is fairly "long" on the right whisker, showing one or more unusually high values (but not quite outliers). The other quartiles are similar in length, showing a fairly even distribution of data.

INTERQUARTILE RANGE

The **interquartile range, or IQR**, is the difference between the upper and lower quartiles. It measures how the data is dispersed: a high IQR means that the data is more spread out, while a low IQR means that the data is clustered more tightly around the median. To find the IQR, subtract the lower quartile value (Q_1) from the upper quartile value (Q_3).

<u>EXAMPLE</u>

To find the upper and lower quartiles, we first find the median and then take the median of all values above it and all values below it. In the following data set (16, 18, 13, 24, 16, 51, 32, 21, 27, 39), we first rearrange the values in numerical order: 13, 16, 16, 18, 21, 24, 27, 32, 39, 51. There are 10 values, so the median is the average of the 5th and 6th: $\frac{21+24}{2} = \frac{45}{2} = 22.5$. We do not actually need this value to find the upper and lower quartiles. We look at the set of numbers below the median: 13, 16, 16, 18, 21. There are five values, so the 3rd is the median (16), or the value of the lower quartile (Q_1). Then we look at the numbers above the median: 24, 27, 32, 39, 51. Again there are five values, so the 3rd is the median (32), or the value of the upper quartile (Q_3). We find the IQR by subtracting Q_1 from Q_3: $32 - 16 = 16$.

68-95-99.7 RULE

The **68–95–99.7 rule** describes how a normal distribution of data should appear when compared to the mean. This is also a description of a normal bell curve. According to this rule, 68 percent of the data values in a normally distributed set should fall within one standard deviation of the mean (34 percent above and 34 percent below the mean), 95 percent of the data values should fall within two standard deviations of the mean (47.5 percent above and 47.5 percent below the mean), and 99.7 percent of the data values should fall within three standard deviations of the mean, again, equally distributed on either side of the mean. This means that only 0.3 percent of all data values should fall more than three standard deviations from the mean. On the graph below, the normal curve is centered on the y-axis. The x-axis labels are how many standard deviations away from the center you are. Therefore, it is easy to see how the 68-95-99.7 rule can apply.

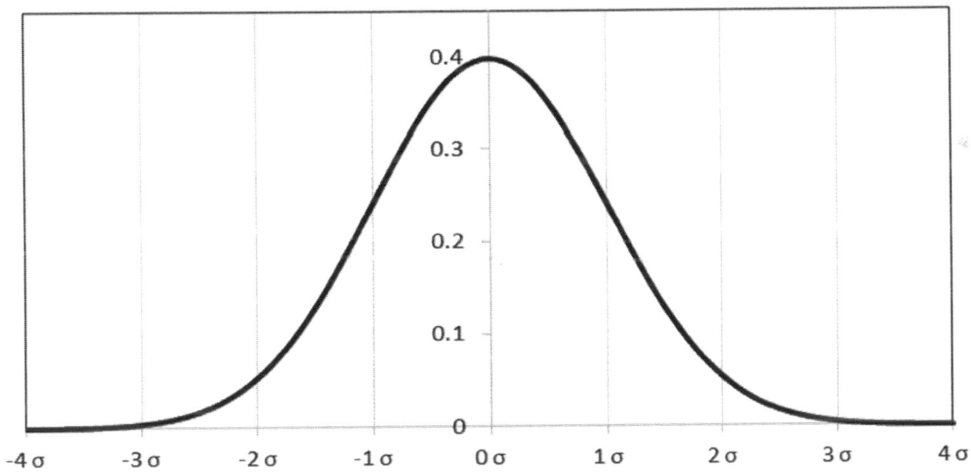

SCATTER PLOTS

BIVARIATE DATA

Bivariate data is simply data from two different variables. (The prefix *bi-* means *two.*) In a *scatter plot*, each value in the set of data is plotted on a grid similar to a Cartesian plane, where each axis represents one of the two variables. By looking at the pattern formed by the points on the grid, you can often determine whether or not there is a relationship between the two variables, and what that relationship is, if it exists. The variables may be directly proportionate, inversely proportionate, or show no proportion at all. It may also be possible to determine if the data is linear, and if so, to find an equation to relate the two variables. The following scatter plot shows the relationship between preference for brand "A" and the age of the consumers surveyed.

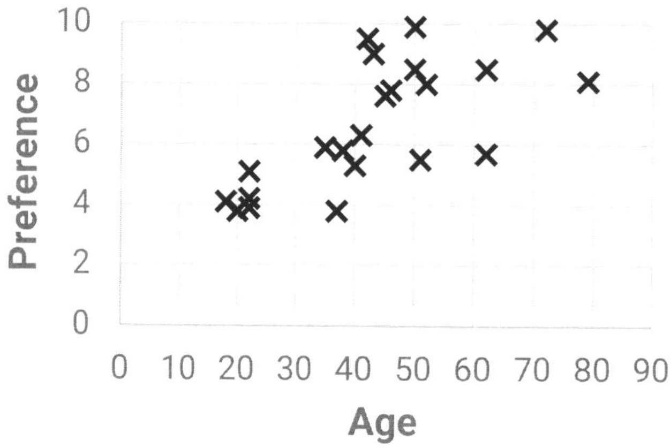

SCATTER PLOTS

Scatter plots are also useful in determining the type of function represented by the data and finding the simple regression. Linear scatter plots may be positive or negative. Nonlinear scatter plots are generally exponential or quadratic. Below are some common types of scatter plots:

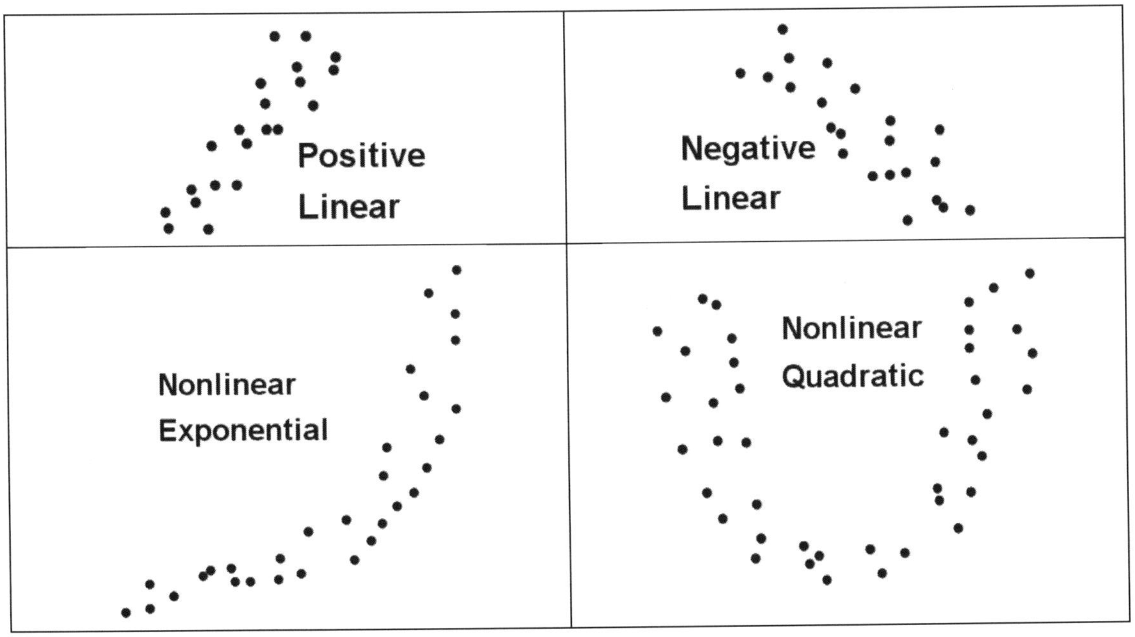

Review Video: Scatter Plot
Visit mometrix.com/academy and enter code: 596526

Statistics

INTRODUCTION TO STATISTICS

Statistics is the branch of mathematics that deals with collecting, recording, interpreting, illustrating, and analyzing large amounts of **data**. The following terms are often used in the discussion of data and **statistics**:

- **Data** – the collective name for pieces of information (singular is datum)
- **Quantitative data** – measurements (such as length, mass, and speed) that provide information about quantities in numbers
- **Qualitative data** – information (such as colors, scents, tastes, and shapes) that cannot be measured using numbers
- **Discrete data** – information that can be expressed only by a specific value, such as whole or half numbers. (e.g., since people can be counted only in whole numbers, a population count would be discrete data.)
- **Continuous data** – information (such as time and temperature) that can be expressed by any value within a given range
- **Primary data** – information that has been collected directly from a survey, investigation, or experiment, such as a questionnaire or the recording of daily temperatures. (Primary data that has not yet been organized or analyzed is called **raw data**.)
- **Secondary data** – information that has been collected, sorted, and processed by the researcher

- **Ordinal data** – information that can be placed in numerical order, such as age or weight
- **Nominal data** – information that *cannot* be placed in numerical order, such as names or places

Data Collection

Population

In statistics, the **population** is the entire collection of people, plants, etc., that data can be collected from. For example, a study to determine how well students in local schools perform on a standardized test would have a population of all the students enrolled in those schools, although a study may include just a small sample of students from each school. A **parameter** is a numerical value that gives information about the population, such as the mean, median, mode, or standard deviation. Remember that the symbol for the mean of a population is μ and the symbol for the standard deviation of a population is σ.

Sample

A **sample** is a portion of the entire population. Whereas a parameter helped describe the population, a **statistic** is a numerical value that gives information about the sample, such as mean, median, mode, or standard deviation. Keep in mind that the symbols for mean and standard deviation are different when they are referring to a sample rather than the entire population. For a sample, the symbol for mean is $\bar{x}$ and the symbol for standard deviation is s. The mean and standard deviation of a sample may or may not be identical to that of the entire population due to a sample only being a subset of the population. However, if the sample is random and large enough, statistically significant values can be attained. Samples are generally used when the population is too large to justify including every element or when acquiring data for the entire population is impossible.

Inferential Statistics

Inferential statistics is the branch of statistics that uses samples to make predictions about an entire population. This type of statistic is often seen in political polls, where a sample of the population is questioned about a particular topic or politician to gain an understanding of the attitudes of the entire population of the country. Often, exit polls are conducted on election days using this method. Inferential statistics can have a large margin of error if you do not have a valid sample.

Sampling Distribution

Statistical values calculated from various samples of the same size make up the **sampling distribution**. For example, if several samples of identical size are randomly selected from a large population and then the mean of each sample is calculated, the distribution of values of the means would be a sampling distribution.

The **sampling distribution of the mean** is the distribution of the sample mean, $\bar{x}$, derived from random samples of a given size. It has three important characteristics. First, the mean of the sampling distribution of the mean is equal to the mean of the population that was sampled. Second, assuming the standard deviation is non-zero, the standard deviation of the sampling distribution of the mean equals the standard deviation of the sampled population divided by the square root of the sample size. This is sometimes called the standard error. Finally, as the sample size gets larger, the sampling distribution of the mean gets closer to a normal distribution via the central limit theorem.

Survey Study

A **survey study** is a method of gathering information from a small group in an attempt to gain enough information to make accurate general assumptions about the population. Once a survey study is completed, the results are then put into a summary report.

Survey studies are generally in the format of surveys, interviews, or questionnaires as part of an effort to find opinions of a particular group or to find facts about a group.

It is important to note that the findings from a survey study are only as accurate as the sample chosen from the population.

Correlational Studies

Correlational studies seek to determine how much one variable is affected by changes in a second variable. For example, correlational studies may look for a relationship between the amount of time a student spends studying for a test and the grade that student earned on the test or between student scores on college admissions tests and student grades in college.

It is important to note that correlational studies cannot show a cause and effect, but rather can show only that two variables are or are not potentially correlated.

Experimental Studies

Experimental studies take correlational studies one step farther, in that they attempt to prove or disprove a cause-and-effect relationship. These studies are performed by conducting a series of experiments to test the hypothesis. For a study to be scientifically accurate, it must have both an experimental group that receives the specified treatment and a control group that does not get the treatment. This is the type of study pharmaceutical companies do as part of drug trials for new medications. Experimental studies are only valid when the proper scientific method has been followed. In other words, the experiment must be well-planned and executed without bias in the testing process, all subjects must be selected at random, and the process of determining which subject is in which of the two groups must also be completely random.

Observational Studies

Observational studies are the opposite of experimental studies. In observational studies, the tester cannot change or in any way control all of the variables in the test. For example, a study to determine which gender does better in math classes in school is strictly observational. You cannot change a person's gender, and you cannot change the subject being studied. The big downfall of observational studies is that you have no way of proving a cause-and-effect relationship because you cannot control outside influences. Events outside of school can influence a student's performance in school, and observational studies cannot take that into consideration.

Random Samples

For most studies, a **random sample** is necessary to produce valid results. Random samples should not have any particular influence to cause sampled subjects to behave one way or another. The goal is for the random sample to be a **representative sample**, or a sample whose characteristics give an accurate picture of the characteristics of the entire population. To accomplish this, you must make sure you have a proper **sample size**, or an appropriate number of elements in the sample.

Biases

In statistical studies, biases must be avoided. **Bias** is an error that causes the study to favor one set of results over another. For example, if a survey to determine how the country views the president's job performance only speaks to registered voters in the president's party, the results will be skewed because a disproportionately large number of responders would tend to show approval, while a disproportionately large number of people in the opposite party would tend to express disapproval. **Extraneous variables** are, as the name implies, outside influences that can affect the outcome of a study. They are not always avoidable but could trigger bias in the result.

DATA ANALYSIS

DISPERSION

A **measure of dispersion** is a single value that helps to "interpret" the measure of central tendency by providing more information about how the data values in the set are distributed about the measure of central tendency. The measure of dispersion helps to eliminate or reduce the disadvantages of using the mean, median, or mode as a single measure of central tendency, and give a more accurate picture of the dataset as a whole. To have a measure of dispersion, you must know or calculate the range, standard deviation, or variance of the data set.

RANGE

The **range** of a set of data is the difference between the greatest and lowest values of the data in the set. To calculate the range, you must first make sure the units for all data values are the same, and then identify the greatest and lowest values. If there are multiple data values that are equal for the highest or lowest, just use one of the values in the formula. Write the answer with the same units as the data values you used to do the calculations.

Review Video: Statistical Range
Visit mometrix.com/academy and enter code: 778541

SAMPLE STANDARD DEVIATION

Standard deviation is a measure of dispersion that compares all the data values in the set to the mean of the set to give a more accurate picture. To find the **standard deviation of a sample**, use the formula

$$s = \sqrt{\frac{\sum_{i=1}^{n}(x_i - \bar{x})^2}{n-1}}$$

Note that s is the standard deviation of a sample, x_i represents the individual values in the data set, $\bar{x}$ is the mean of the data values in the set, and n is the number of data values in the set. The higher the value of the standard deviation is, the greater the variance of the data values from the mean. The units associated with the standard deviation are the same as the units of the data values.

Review Video: Standard Deviation
Visit mometrix.com/academy and enter code: 419469

SAMPLE VARIANCE

The **variance of a sample** is the square of the sample standard deviation (denoted s^2). While the mean of a set of data gives the average of the set and gives information about where a specific data value lies in relation to the average, the variance of the sample gives information about the degree to which the data values are spread out and tells you how close an individual value is to the average compared to the other values. The units associated with variance are the same as the units of the data values squared.

Percentile

Percentiles and quartiles are other methods of describing data within a set. **Percentiles** tell what percentage of the data in the set fall below a specific point. For example, achievement test scores are often given in percentiles. A score at the 80th percentile is one which is equal to or higher than 80 percent of the scores in the set. In other words, 80 percent of the scores were lower than that score.

Quartiles are percentile groups that make up quarter sections of the data set. The first quartile is the 25th percentile. The second quartile is the 50th percentile; this is also the median of the dataset. The third quartile is the 75th percentile.

Skewness

Skewness is a way to describe the symmetry or asymmetry of the distribution of values in a dataset. If the distribution of values is symmetrical, there is no skew. In general the closer the mean of a data set is to the median of the data set, the less skew there is. Generally, if the mean is to the right of the median, the data set is *positively skewed*, or right-skewed, and if the mean is to the left of the median, the data set is *negatively skewed*, or left-skewed. However, this rule of thumb is not infallible. When the data values are graphed on a curve, a set with no skew will be a perfect bell curve.

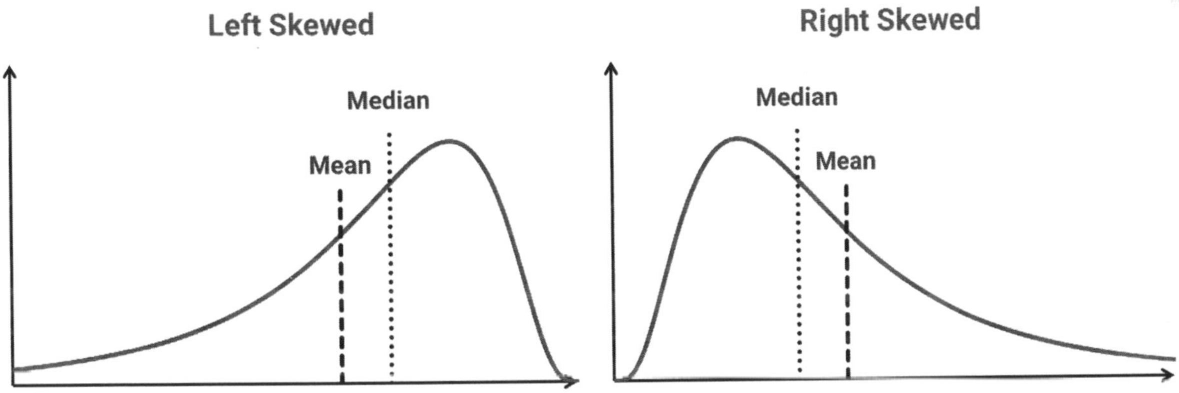

To estimate skew, use the formula:

$$\text{skew} = \frac{\sqrt{n(n-1)}}{n-2}\left(\frac{\frac{1}{n}\sum_{i=1}^{n}(x_i-\bar{x})^3}{\left(\frac{1}{n}\sum_{i=1}^{n}(x_i-\bar{x})^2\right)^{\frac{3}{2}}}\right)$$

Note that n is the datapoints in the set, x_i is the ith value in the set, and $\bar{x}$ is the mean of the set.

Review Video: Skew
Visit mometrix.com/academy and enter code: 661486

Unimodal vs. Bimodal

If a distribution has a single peak, it would be considered **unimodal**. If it has two discernible peaks it would be considered **bimodal**. Bimodal distributions may be an indication that the set of data being considered is actually the combination of two sets of data with significant differences. A

uniform distribution is a distribution in which there is *no distinct peak or variation* in the data. No values or ranges are particularly more common than any other values or ranges.

Outlier

An outlier is an extremely high or extremely low value in the data set. It may be the result of measurement error, in which case, the outlier is not a valid member of the data set. However, it may also be a valid member of the distribution. Unless a measurement error is identified, the experimenter cannot know for certain if an outlier is or is not a member of the distribution. There are arbitrary methods that can be employed to designate an extreme value as an outlier. One method designates an outlier (or possible outlier) to be any value less than $Q_1 - 1.5(IQR)$ or any value greater than $Q_3 + 1.5(IQR)$.

Data Analysis

Simple Regression

In statistics, **simple regression** is using an equation to represent a relation between independent and dependent variables. The independent variable is also referred to as the explanatory variable or the predictor and is generally represented by the variable x in the equation. The dependent variable, usually represented by the variable y, is also referred to as the response variable. The equation may be any type of function – linear, quadratic, exponential, etc. The best way to handle this task is to use the regression feature of your graphing calculator. This will easily give you the curve of best fit and provide you with the coefficients and other information you need to derive an equation.

Line of Best Fit

In a scatter plot, the **line of best fit** is the line that best shows the trends of the data. The line of best fit is given by the equation $\hat{y} = ax + b$, where a and b are the regression coefficients. The regression coefficient a is also the slope of the line of best fit, and b is also the y-coordinate of the point at which the line of best fit crosses the y-axis. Not every point on the scatter plot will be on the line of best fit. The differences between the y-values of the points in the scatter plot and the corresponding y-values according to the equation of the line of best fit are the residuals. The line of best fit is also called the least-squares regression line because it is also the line that has the lowest sum of the squares of the residuals.

Correlation Coefficient

The **correlation coefficient** is the numerical value that indicates how strong the relationship is between the two variables of a linear regression equation. A correlation coefficient of –1 is a perfect negative correlation. A correlation coefficient of +1 is a perfect positive correlation. Correlation coefficients close to –1 or +1 are very strong correlations. A correlation coefficient equal to zero indicates there is no correlation between the two variables. This test is a good indicator of whether or not the equation for the line of best fit is accurate. The formula for the correlation coefficient is

$$r = \frac{\sum_{i=1}^{n}(x_i - \bar{x})(y_i - \bar{y})}{\sqrt{\sum_{i=1}^{n}(x_i - \bar{x})^2}\sqrt{\sum_{i=1}^{n}(y_i - \bar{y})^2}}$$

where r is the correlation coefficient, n is the number of data values in the set, (x_i, y_i) is a point in the set, and $\bar{x}$ and $\bar{y}$ are the means.

Z-Score

A **z-score** is an indication of how many standard deviations a given value falls from the sample mean. To calculate a z-score, use the formula:

$$\frac{x - \bar{x}}{\sigma}$$

In this formula x is the data value, $\bar{x}$ is the mean of the sample data, and σ is the standard deviation of the population. If the z-score is positive, the data value lies above the mean. If the z-score is negative, the data value falls below the mean. These scores are useful in interpreting data such as standardized test scores, where every piece of data in the set has been counted, rather than just a small random sample. In cases where standard deviations are calculated from a random sample of the set, the z-scores will not be as accurate.

Central Limit Theorem

According to the **central limit theorem**, regardless of what the original distribution of a sample is, the distribution of the means tends to get closer and closer to a normal distribution as the sample size gets larger and larger (this is necessary because the sample is becoming more all-encompassing of the elements of the population). As the sample size gets larger, the distribution of the sample mean will approach a normal distribution with a mean of the population mean and a variance of the population variance divided by the sample size.

Measures of Central Tendency

A **measure of central tendency** is a statistical value that gives a reasonable estimate for the center of a group of data. There are several different ways of describing the measure of central tendency. Each one has a unique way it is calculated, and each one gives a slightly different perspective on the data set. Whenever you give a measure of central tendency, always make sure the units are the same. If the data has different units, such as hours, minutes, and seconds, convert all the data to the same unit, and use the same unit in the measure of central tendency. If no units are given in the data, do not give units for the measure of central tendency.

Mean

The **statistical mean** of a group of data is the same as the arithmetic average of that group. To find the mean of a set of data, first convert each value to the same units, if necessary. Then find the sum of all the values, and count the total number of data values, making sure you take into consideration each individual value. If a value appears more than once, count it more than once. Divide the sum of the values by the total number of values and apply the units, if any. Note that the mean does not have to be one of the data values in the set, and may not divide evenly.

$$\text{mean} = \frac{\text{sum of the data values}}{\text{quantity of data values}}$$

For instance, the mean of the data set {88, 72, 61, 90, 97, 68, 88, 79, 86, 93, 97, 71, 80, 84, 89} would be the sum of the fifteen numbers divided by 15:

$$\frac{88 + 72 + 61 + 90 + 97 + 68 + 88 + 79 + 86 + 93 + 97 + 71 + 80 + 84 + 89}{15} = \frac{1242}{15} = 82.8$$

While the mean is relatively easy to calculate and averages are understood by most people, the mean can be very misleading if it is used as the sole measure of central tendency. If the data set has outliers (data values that are unusually high or unusually low compared to the rest of the data values), the mean can be very distorted, especially if the data set has a small number of values. If unusually high values are countered with unusually low values, the mean is not affected as much. For example, if five of twenty students in a class get a 100 on a test, but the other 15 students have an average of 60 on the same test, the class average would appear as 70. Whenever the mean is skewed by outliers, it is always a good idea to include the median as an alternate measure of central tendency.

A **weighted mean**, or weighted average, is a mean that uses "weighted" values. The formula is weighted mean $= \frac{w_1x_1+w_2x_2+w_3x_3...+w_nx_n}{w_1+w_2+w_3+\cdots+w_n}$. Weighted values, such as $w_1, w_2, w_3, ... w_n$ are assigned to each member of the set $x_1, x_2, x_3, ... x_n$. When calculating the weighted mean, make sure a weight value for each member of the set is used.

Review Video: All About Averages
Visit mometrix.com/academy and enter code: 176521

Median

The **statistical median** is the value in the middle of the set of data. To find the median, list all data values in order from smallest to largest or from largest to smallest. Any value that is repeated in the set must be listed the number of times it appears. If there are an odd number of data values, the median is the value in the middle of the list. If there is an even number of data values, the median is the arithmetic mean of the two middle values.

For example, the median of the data set {88, 72, 61, 90, 97, 68, 88, 79, 86, 93, 97, 71, 80, 84, 88} is 86 since the ordered set is {61, 68, 71, 72, 79, 80, 84, **86**, 88, 88, 88, 90, 93, 97, 97}.

The big disadvantage of using the median as a measure of central tendency is that is relies solely on a value's relative size as compared to the other values in the set. When the individual values in a set of data are evenly dispersed, the median can be an accurate tool. However, if there is a group of rather large values or a group of rather small values that are not offset by a different group of values, the information that can be inferred from the median may not be accurate because the distribution of values is skewed.

Mode

The **statistical mode** is the data value that occurs the greatest number of times in the data set. It is possible to have exactly one mode, more than one mode, or no mode. To find the mode of a set of data, arrange the data like you do to find the median (all values in order, listing all multiples of data values). Count the number of times each value appears in the data set. If all values appear an equal number of times, there is no mode. If one value appears more than any other value, that value is the mode. If two or more values appear the same number of times, but there are other values that appear fewer times and no values that appear more times, all of those values are the modes.

For example, the mode of the data set {**88**, 72, 61, 90, 97, 68, **88**, 79, 86, 93, 97, 71, 80, 84, **88**} is 88.

The main disadvantage of the mode is that the values of the other data in the set have no bearing on the mode. The mode may be the largest value, the smallest value, or a value anywhere in between in the set. The mode only tells which value or values, if any, occurred the greatest number of times. It does not give any suggestions about the remaining values in the set.

Review Video: Mean, Median, and Mode
Visit mometrix.com/academy and enter code: 286207

Probability

Probability

Probability is the likelihood of a certain outcome occurring for a given event. An **event** is any situation that produces a result. It could be something as simple as flipping a coin or as complex as launching a rocket. Determining the probability of an outcome for an event can be equally simple or complex. As such, there are specific terms used in the study of probability that need to be understood:

- **Compound event**—an event that involves two or more independent events (rolling a pair of dice and taking the sum)
- **Desired outcome** (or success)—an outcome that meets a particular set of criteria (a roll of 1 or 2 if we are looking for numbers less than 3)
- **Independent events**—two or more events whose outcomes do not affect one another (two coins tossed at the same time)
- **Dependent events**—two or more events whose outcomes affect one another (two cards drawn consecutively from the same deck)
- **Certain outcome**—probability of outcome is 100% or 1
- **Impossible outcome**—probability of outcome is 0% or 0
- **Mutually exclusive outcomes**—two or more outcomes whose criteria cannot all be satisfied in a single event (a coin coming up heads and tails on the same toss)
- **Random variable**—refers to all possible outcomes of a single event which may be discrete or continuous.

Review Video: Intro to Probability
Visit mometrix.com/academy and enter code: 212374

Sample Space

The total set of all possible results of a test or experiment is called a **sample space**, or sometimes a universal sample space. The sample space, represented by one of the variables S, Ω, or U (for universal sample space) has individual elements called outcomes. Other terms for outcome that may be used interchangeably include elementary outcome, simple event, or sample point. The number of outcomes in a given sample space could be infinite or finite, and some tests may yield multiple unique sample sets. For example, tests conducted by drawing playing cards from a standard deck would have one sample space of the card values, another sample space of the card suits, and a third sample space of suit-denomination combinations. For most tests, the sample spaces considered will be finite.

An **event**, represented by the variable E, is a portion of a sample space. It may be one outcome or a group of outcomes from the same sample space. If an event occurs, then the test or experiment will generate an outcome that satisfies the requirement of that event. For example, given a standard deck of 52 playing cards as the sample space, and defining the event as the collection of face cards, then the event will occur if the card drawn is a J, Q, or K. If any other card is drawn, the event is said to have not occurred.

For every sample space, each possible outcome has a specific likelihood, or probability, that it will occur. The probability measure, also called the **distribution**, is a function that assigns a real number probability, from zero to one, to each outcome. For a probability measure to be accurate, every outcome must have a real number probability measure that is greater than or equal to zero and less than or equal to one. Also, the probability measure of the sample space must equal one, and the probability measure of the union of multiple outcomes must equal the sum of the individual probability measures.

Probabilities of events are expressed as real numbers from zero to one. They give a numerical value to the chance that a particular event will occur. The probability of an event occurring is the sum of the probabilities of the individual elements of that event. For example, in a standard deck of 52 playing cards as the sample space and the collection of face cards as the event, the probability of drawing a specific face card is $\frac{1}{52} = 0.019$, but the probability of drawing any one of the twelve face cards is $12(0.019) = 0.228$. Note that rounding of numbers can generate different results. If you multiplied 12 by the fraction $\frac{1}{52}$ before converting to a decimal, you would get the answer $\frac{12}{52} = 0.231$.

Theoretical and Experimental Probability

Theoretical probability can usually be determined without actually performing the event. The likelihood of an outcome occurring, or the probability of an outcome occurring, is given by the formula:

$$P(A) = \frac{\text{Number of acceptable outcomes}}{\text{Number of possible outcomes}}$$

Note that $P(A)$ is the probability of an outcome A occurring, and each outcome is just as likely to occur as any other outcome. If each outcome has the same probability of occurring as every other possible outcome, the outcomes are said to be equally likely to occur. The total number of acceptable outcomes must be less than or equal to the total number of possible outcomes. If the two are equal, then the outcome is certain to occur and the probability is 1. If the number of acceptable outcomes is zero, then the outcome is impossible and the probability is 0. For example, if there are 20 marbles in a bag and 5 are red, then the theoretical probability of randomly selecting a red marble is 5 out of 20, $\left(\frac{5}{20} = \frac{1}{4}, 0.25, \text{or } 25\%\right)$.

If the theoretical probability is unknown or too complicated to calculate, it can be estimated by an experimental probability. **Experimental probability**, also called empirical probability, is an estimate of the likelihood of a certain outcome based on repeated experiments or collected data. In other words, while theoretical probability is based on what *should* happen, experimental probability is based on what *has* happened. Experimental probability is calculated in the same way as theoretical probability, except that actual outcomes are used instead of possible outcomes. The more experiments performed or datapoints gathered, the better the estimate should be.

Theoretical and experimental probability do not always line up with one another. Theoretical probability says that out of 20 coin-tosses, 10 should be heads. However, if we were actually to toss 20 coins, we might record just 5 heads. This doesn't mean that our theoretical probability is incorrect; it just means that this particular experiment had results that were different from what was predicted. A practical application of empirical probability is the insurance industry. There are no set functions that define lifespan, health, or safety. Insurance companies look at factors from hundreds of thousands of individuals to find patterns that they then use to set the formulas for insurance premiums.

Review Video: Empirical Probability
Visit mometrix.com/academy and enter code: 513468

Objective and Subjective Probability

Objective probability is based on mathematical formulas and documented evidence. Examples of objective probability include raffles or lottery drawings where there is a pre-determined number of possible outcomes and a predetermined number of outcomes that correspond to an event. Other cases of objective probability include probabilities of rolling dice, flipping coins, or drawing cards. Most gambling games are based on objective probability.

In contrast, **subjective probability** is based on personal or professional feelings and judgments. Often, there is a lot of guesswork following extensive research. Areas where subjective probability is applicable include sales trends and business expenses. Attractions set admission prices based on subjective probabilities of attendance based on varying admission rates in an effort to maximize their profit.

Complement of an Event

Sometimes it may be easier to calculate the possibility of something not happening, or the **complement of an event**. Represented by the symbol $\bar{A}$, the complement of A is the probability that event A does not happen. When you know the probability of event A occurring, you can use the formula $P(\bar{A}) = 1 - P(A)$, where $P(\bar{A})$ is the probability of event A not occurring, and $P(A)$ is the probability of event A occurring.

Addition Rule

The **addition rule** for probability is used for finding the probability of a compound event. Use the formula $P(A \cup B) = P(A) + P(B) - P(A \cap B)$, where $P(A \cap B)$ is the probability of both events occurring to find the probability of a compound event. The probability of both events occurring at the same time must be subtracted to eliminate any overlap in the first two probabilities.

Conditional Probability

Given two events A and B, the **conditional probability** $P(A|B)$ is the probability that event A will occur, given that event B has occurred. The conditional probability cannot be calculated simply from $P(A)$ and $P(B)$; these probabilities alone do not give sufficient information to determine the conditional probability. It can, however, be determined if you are also given the probability of the intersection of events A and B, $P(A \cap B)$, the probability that events A and B both occur. Specifically, $P(A|B) = \frac{P(A\cap B)}{P(B)}$. For instance, suppose you have a jar containing two red marbles and two blue marbles, and you draw two marbles at random. Consider event A being the event that the first marble drawn is red, and event B being the event that the second marble drawn is blue. If we want to find the probability that B occurs given that A occurred, $P(B|A)$, then we can compute it using the fact that $P(A)$ is $\frac{1}{2}$, and $P(A \cap B)$ is $\frac{1}{3}$. (The latter may not be obvious, but may be determined by finding the product of $\frac{1}{2}$ and $\frac{2}{3}$). Therefore $P(B|A) = \frac{P(A\cap B)}{P(A)} = \frac{1/3}{1/2} = \frac{2}{3}$.

Conditional Probability in Everyday Situations

Conditional probability often arises in everyday situations in, for example, estimating the risk or benefit of certain activities. The conditional probability of having a heart attack given that you exercise daily may be smaller than the overall probability of having a heart attack. The conditional probability of having lung cancer given that you are a smoker is larger than the overall probability of having lung cancer. Note that changing the order of the conditional probability changes the meaning: the conditional probability of having lung cancer given that you are a smoker is a very different thing from the probability of being a smoker given that you have lung cancer. In an extreme case, suppose that a certain rare disease is caused only by eating a certain food, but even then, it is unlikely. Then the conditional probability of having that disease given that you eat the dangerous food is nonzero but low, but the conditional probability of having eaten that food given that you have the disease is 100%!

Review Video: Conditional Probability
Visit mometrix.com/academy and enter code: 397924

Independence

The conditional probability $P(A|B)$ is the probability that event A will occur given that event B occurs. If the two events are independent, we do not expect that whether or not event B occurs should have any effect on whether or not event A occurs. In other words, we expect $P(A|B) = P(A)$.

This can be proven using the usual equations for conditional probability and the joint probability of independent events. The conditional probability $P(A|B) = \frac{P(A \cap B)}{P(B)}$. If A and B are independent, then $P(A \cap B) = P(A)P(B)$. So $P(A|B) = \frac{P(A)P(B)}{P(B)} = P(A)$. By similar reasoning, if A and B are independent then $P(B|A) = P(B)$.

Multiplication Rule

The **multiplication rule** can be used to find the probability of two independent events occurring using the formula $P(A \cap B) = P(A) \times P(B)$, where $P(A \cap B)$ is the probability of two independent events occurring, $P(A)$ is the probability of the first event occurring, and $P(B)$ is the probability of the second event occurring.

The multiplication rule can also be used to find the probability of two dependent events occurring using the formula $P(A \cap B) = P(A) \times P(B|A)$, where $P(A \cap B)$ is the probability of two dependent events occurring and $P(B|A)$ is the probability of the second event occurring after the first event has already occurred.

Use a **combination of the multiplication** rule and the rule of complements to find the probability that at least one outcome of the element will occur. This is given by the general formula $P(\text{at least one event occurring}) = 1 - P(\text{no outcomes occurring})$. For example, to find the probability that at least one even number will show when a pair of dice is rolled, find the probability that two odd numbers will be rolled (no even numbers) and subtract from one. You can always use a tree diagram or make a chart to list the possible outcomes when the sample space is small, such as in the dice-rolling example, but in most cases it will be much faster to use the multiplication and complement formulas.

Review Video: Multiplication Rule
Visit mometrix.com/academy and enter code: 782598

Union and Intersection of Two Sets of Outcomes

If A and B are each a set of elements or outcomes from an experiment, then the **union** (symbol ∪) of the two sets is the set of elements found in set A or set B. For example, if $A = \{2, 3, 4\}$ and $B = \{3, 4, 5\}$, $A \cup B = \{2, 3, 4, 5\}$. Note that the outcomes 3 and 4 appear only once in the union. For statistical events, the union is equivalent to "or"; $P(A \cup B)$ is the same thing as $P(A \text{ or } B)$. The **intersection** (symbol ∩) of two sets is the set of outcomes common to both sets. For the above sets

A and B, $A \cap B = \{3, 4\}$. For statistical events, the intersection is equivalent to "and"; $P(A \cap B)$ is the same thing as $P(A \text{ and } B)$. It is important to note that union and intersection operations commute. That is:

$$A \cup B = B \cup A \text{ and } A \cap B = B \cap A$$

Permutations and Combinations in Probability

When trying to calculate the probability of an event using the $\frac{\text{desired outcomes}}{\text{total outcomes}}$ formula, you may frequently find that there are too many outcomes to individually count them. **Permutation** and **combination formulas** offer a shortcut to counting outcomes. A permutation is an arrangement of a specific number of a set of objects in a specific order. The number of **permutations** of r items given a set of n items can be calculated as ${}_nP_r = \frac{n!}{(n-r)!}$. Combinations are similar to permutations, except there are no restrictions regarding the order of the elements. While ABC is considered a different permutation than BCA, ABC and BCA are considered the same combination. The number of **combinations** of r items given a set of n items can be calculated as ${}_nC_r = \frac{n!}{r!(n-r)!}$ or ${}_nC_r = \frac{{}_nP_r}{r!}$.

Suppose you want to calculate how many different 5-card hands can be drawn from a deck of 52 cards. This is a combination since the order of the cards in a hand does not matter. There are 52 cards available, and 5 to be selected. Thus, the number of different hands is ${}_{52}C_5 = \frac{52!}{5! \times 47!} =$ 2,598,960.

Review Video: Probability - Permutation and Combination
Visit mometrix.com/academy and enter code: 907664

Tree Diagrams

For a simple sample space, possible outcomes may be determined by using a **tree diagram** or an organized chart. In either case, you can easily draw or list out the possible outcomes. For example, to determine all the possible ways three objects can be ordered, you can draw a tree diagram:

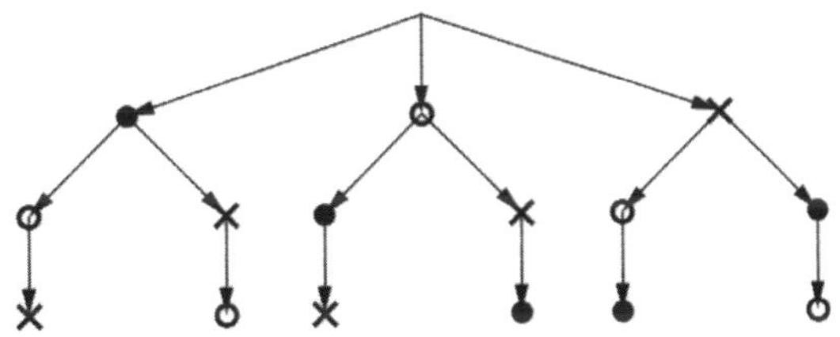

Review Video: Tree Diagrams
Visit mometrix.com/academy and enter code: 829158

You can also make a chart to list all the possibilities:

First object	Second object	Third object
●	x	o
●	o	x
o	●	x
o	x	●
x	●	o
x	o	●

Either way, you can easily see there are six possible ways the three objects can be ordered.

If two events have no outcomes in common, they are said to be **mutually exclusive**. For example, in a standard deck of 52 playing cards, the event of all card suits is mutually exclusive to the event of all card values. If two events have no bearing on each other so that one event occurring has no influence on the probability of another event occurring, the two events are said to be independent. For example, rolling a standard six-sided die multiple times does not change that probability that a particular number will be rolled from one roll to the next. If the outcome of one event does affect the probability of the second event, the two events are said to be dependent. For example, if cards are drawn from a deck, the probability of drawing an ace after an ace has been drawn is different than the probability of drawing an ace if no ace (or no other card, for that matter) has been drawn.

In probability, the **odds in favor of an event** are the number of times the event will occur compared to the number of times the event will not occur. To calculate the odds in favor of an event, use the formula $\frac{P(A)}{1-P(A)}$, where $P(A)$ is the probability that the event will occur. Many times, odds in favor is given as a ratio in the form $\frac{a}{b}$ or $a{:}\,b$, where a is the probability of the event occurring and b is the complement of the event, the probability of the event not occurring. If the odds in favor are given as 2:5, that means that you can expect the event to occur two times for every 5 times that it does not occur. In other words, the probability that the event will occur is $\frac{2}{2+5} = \frac{2}{7}$.

In probability, the **odds against an event** are the number of times the event will not occur compared to the number of times the event will occur. To calculate the odds against an event, use the formula $\frac{1-P(A)}{P(A)}$, where $P(A)$ is the probability that the event will occur. Many times, odds against is given as a ratio in the form $\frac{b}{a}$ or $b{:}\,a$, where b is the probability the event will not occur (the complement of the event) and a is the probability the event will occur. If the odds against an event are given as 3:1, that means that you can expect the event to not occur 3 times for every one time it does occur. In other words, 3 out of every 4 trials will fail.

Two-Way Frequency Tables

If we have a two-way frequency table, it is generally a straightforward matter to read off the probabilities of any two events A and B, as well as the joint probability of both events occurring, $P(A \cap B)$. We can then find the conditional probability $P(A|B)$ by calculating $P(A|B) = \frac{P(A \cap B)}{P(B)}$. We could also check whether or not events are independent by verifying whether $P(A)P(B) = P(A \cap B)$.

For example, a certain store's recent T-shirt sales:

	Small	Medium	Large	Total
Blue	25	40	35	100
White	27	25	22	74
Black	8	23	15	46
Total	60	88	72	220

Suppose we want to find the conditional probability that a customer buys a black shirt (event A), given that the shirt he buys is size small (event B). From the table, the probability $P(B)$ that a customer buys a small shirt is $\frac{60}{220} = \frac{3}{11}$. The probability $P(A \cap B)$ that he buys a small, black shirt is $\frac{8}{220} = \frac{2}{55}$. The conditional probability $P(A|B)$ that he buys a black shirt, given that he buys a small shirt, is therefore $P(A|B) = \frac{2/55}{3/11} = \frac{2}{15}$.

Similarly, if we want to check whether the event a customer buys a blue shirt, A, is independent of the event that a customer buys a medium shirt, B. From the table, $P(A) = \frac{100}{220} = \frac{5}{11}$ and $P(B) = \frac{88}{220} = \frac{4}{10}$. Also, $P(A \cap B) = \frac{40}{220} = \frac{2}{11}$. Since $\left(\frac{5}{11}\right)\left(\frac{4}{10}\right) = \frac{20}{110} = \frac{2}{11}$, $P(A)P(B) = P(A \cap B)$ and these two events are indeed independent.

Expected Value

Expected value is a method of determining the expected outcome in a random situation. It is a sum of the weighted probabilities of the possible outcomes. Multiply the probability of an event occurring by the weight assigned to that probability (such as the amount of money won or lost). A practical application of the expected value is to determine whether a game of chance is really fair. If the sum of the weighted probabilities is equal to zero, the game is generally considered fair because the player has a fair chance to at least break even. If the expected value is less than zero, then players are expected to lose more than they win. For example, a lottery drawing might allow the player to choose any three-digit number, 000–999. The probability of choosing the winning number is 1:1000. If it costs \$1 to play, and a winning number receives \$500, the expected value is $\left(-\$1 \times \frac{999}{1{,}000}\right) + \left(\$499 \times \frac{1}{1{,}000}\right) = -\$0.50$. You can expect to lose on average 50 cents for every dollar you spend.

Review Video: Expected Value
Visit mometrix.com/academy and enter code: 643554

Expected Value and Simulators

A die roll simulator will show the results of n rolls of a die. The result of each die roll may be recorded. For example, suppose a die is rolled 100 times. All results may be recorded. The numbers of 1s, 2s, 3s, 4s, 5s, and 6s, may be counted. The experimental probability of rolling each number will equal the ratio of the frequency of the rolled number to the total number of rolls. As the number of rolls increases, or approaches infinity, the experimental probability will approach the theoretical probability of $\frac{1}{6}$. Thus, the expected value for the roll of a die is shown to be $\left(1 \times \frac{1}{6}\right) + \left(2 \times \frac{1}{6}\right) + \left(3 \times \frac{1}{6}\right) + \left(4 \times \frac{1}{6}\right) + \left(5 \times \frac{1}{6}\right) + \left(6 \times \frac{1}{6}\right)$, or 3.5.

Set Theory

Set Theory

Discrete Mathematics

Among mathematicians, there is not an agreed-upon definition of discrete math. What is agreed upon is the fact that discrete math deals with processes that use a finite, or countable, number of elements. In discrete math, the elements will be discontinuous, as this branch of mathematics does not involve the continuity that processes of calculus do. Generally, discrete math uses countable sets of rational numbers, although they do not use the set of all real numbers, as that would then make the math continuous and put it in the category of algebra or calculus. Discrete math has numerous applications in the fields of computer science and business.

Review Video: Introduction to Set Theory
Visit mometrix.com/academy and enter code: 397496

Element of a Set

A set is a mathematical collection of items, and an element is an item that is included in the set. These items are typically sets of numbers, or sets of geometrical points, or even sets of sets.

Whether an item is an **element of a set** is a binary property. In other words, a particular item either is or is not an element of a given set. There are not different degrees of belonging to a set, and one element of a set cannot be more an element than another. Additionally, the elements do not have any particular order, and there is no count of how many times an element appears in a set. For example, in the set $\{0, 2, 1, 2, 4\}$, the numbers 0, 1, 2, and 4 are elements of the set. Two is not more of an element than the others because it is listed twice. Any number other than these, such as 3, is not an element of the set.

The mathematical symbol $\in$ means "is an element of." for instance, "$1 \in A$" means that "the number 1 is an element of set A". The symbol can be negated with a diagonal slash, so "$1 \notin A$" means "the number 1 is *not* an element of set A."

Empty Set

The empty set contains no elements. The symbol for the **empty set** is a circle with a line through it, $\emptyset$, or the empty set can be written in roster form as $\{\quad\}$. Although it may seem trivial, the empty set is important for the same reason that zero is an important number—without the empty set, many set operations and definitions would be incomplete. For instance, the intersection of two non-overlapping sets is the empty set. The set of all prime numbers that are perfect squares is the empty set, because there are no such numbers. Other sets can be constructed using the empty set. For instance, a set *containing* the empty set, $\{\{\quad\}\}$ or $\{\emptyset\}$, is different from the empty set itself, since it's not actually empty—it contains one element, the empty set.

The empty set is, by definition, a subset of every set, including itself. This is because every element that is an element of the empty set is also an element of any other set, since the empty set contains no elements that could make this statement false.

Subset, Proper Subset, and Superset

Set A is a **subset** of set B if set A is contained entirely within set B. More formally, set A is a subset of set B if every element of set A is also in set B. We can write this as $\forall x(x \in A \Rightarrow x \in B)$. The symbol for subset is $\subseteq$, so we can write "$A \subseteq B$" to mean "set A is a subset of set B," or "$A \nsubseteq B$" to mean "set A is NOT a subset of set B".

Any set is a subset of itself, since naturally any element in a set is in that set. A **proper subset** means a subset that is not equal to the other set. In other words, set A is a proper subset of set B if set A is a subset of set B but they are not the same, so $A \subseteq B$ but $B \nsubseteq A$. For example, $\{0, 1, 2\}$ is a proper subset of $\{0, 1, 2, 3\}$ because every element in the first set is included in the second set but the two sets are not identical. The symbol for a proper subset is $\subset$, so we write "$A \subset B$" to mean "set A is a proper subset of set B," and "$A \not\subset B$" to mean "set A is not a proper subset of set B."

The converse of a subset is a **superset**. If A is a subset of B, then B is a superset of A. In the example above, $\{0, 1, 2, 3\}$ is a superset of $\{0, 1, 2\}$ because the first set contains every element of the second plus more.

Union of Two or More Sets

The **union** of two or more sets includes all the elements that are included in all of the sets. An element is in the union of sets A and B if it is in set A or in set B, or in both. The union is written with the symbol $\cup$, so "$A \cup B$" (read "A union B") means the union of sets A and B.

For example, given the sets $A = \{1, 2, 3, 4\}$ and $B = \{2, 4, 6, 8\}$, then $A \cup B = \{1, 2, 3, 4, 6, 8\}$. 1 and 3 are in $A \cup B$ because they are in set A, and 6 and 8 are in $A \cup B$ because they are in set B. 2 and 4 appear in both A and B, so they are in $A \cup B$, but they only appear once.

The union operation on sets is both commutative and associative. That is, $A \cup B = B \cup A$, and $(A \cup B) \cup C = A \cup (B \cup C)$.

Intersection of Two or More Sets

The intersection of two or more sets is a set including every element that appears in *all* of the sets. An element is in the intersection of sets A and B if and only if, the element appears in both set A and set B. For instance, if set $A = \{0, 1, 2, 3\}$ and set $B = \{2, 3, 4, 5\}$, the intersection of sets A and B includes the elements 2 and 3, because these are the only elements that appear in both sets. The intersection is written with the symbol $\cap$, so "$A \cap B$" (read "A intersection B") means the intersection of sets A and B.

The intersection operation on sets is both commutative and associative. That is, $A \cap B = B \cap A$, and $(A \cap B) \cap C = A \cap (B \cap C)$. The union and intersection operations also jointly have distributive properties: $A \cap (B \cup C) = (A \cap B) \cup (A \cap C)$ and $A \cup (B \cap C) = (A \cup B) \cap (A \cup C)$.

Complement of a Set

The complement of a set is a set containing all the elements that are *not* in the original set. To determine this, we must also define the **universe of discourse**—the set of all possible elements we're considering, abbreviated U. For mathematical applications, the universe of discourse is commonly the set of all integers (abbreviated $\mathbb{Z}$), or the set of all real numbers (abbreviated $\mathbb{R}$). If U is the set of all integers, for example, then the complement of the set $A = \{1, 2, 3\}$ would include elements such as -3, 0, or 6, but it would not include elements such as $\sqrt{7}$ or $\frac{1}{4}$, because these are not in the universe of discourse.

The **complement of a set** is often written either by drawing a line over the name of the original set, or by adding an apostrophe or a superscripted C. The complement of set A, for instance, could be written as $\bar{A}$, as A', or as A^C.

For example, given set $A = \{2, 4, 6, 8\}$ with a universe of discourse of $U = \{1, 2, 3, 4, 5, 6, 7, 8, 9, 10\}$, the complement of set A would be $\{1, 3, 5, 7, 9, 10\}$—all the elements of U that are not in A.

DIFFERENCE OF TWO SETS

The difference of two sets is a set containing all the elements that are in the first set but not in the second. The difference of sets A and B is also referred to as the **relative complement** of set A with respect to set B, and is written $A \setminus B$ or $A - B$. The difference is always a subset of the first set: $A \setminus B \subseteq A$. However, it is *not* necessarily the case that $A \setminus B \subseteq B$. In fact, because any element of $A \setminus B$ is by definition not an element of B, the only time that $A \setminus B$ can be a subset of B is if $A \setminus B$ has no elements, i.e., if $A \setminus B = \emptyset$, which is true if and only if, $A \subseteq B$.

Unlike the union and intersection, the operation of the difference between sets is neither commutative nor associative. In general, $A \setminus B \neq B \setminus A$, and $A \setminus (B \setminus C) \neq (A \setminus B) \setminus C$.

For example, consider the sets $A = \{1, 2, 3\}$, $B = \{2, 4\}$, and $C = \{1, 2\}$. Then $A \setminus B = \{1, 3\}$ since 1 and 3 are in set A but not in set B. Similarly, $B \setminus A = \{4\}$, $B \setminus C = \{4\}$, $C \setminus B = \{1\}$, and $A \setminus C = \{3\}$. However, $C \setminus A = \emptyset$, i.e., set C is a subset of set A, so there are no elements that appear in C but not in A, which means the difference between C and A is the empty set.

SYMBOLS FOR SETS COMMONLY USED IN MATHEMATICS

Certain sets are used frequently enough in mathematics to have their own symbols. The standard symbols for these sets resemble upper-case letters of the Latin alphabet, but in a typeface with double lines. Among the most commonly used **sets** are:

- $\mathbb{R}$ – the set of all real numbers
- $\mathbb{Z}$ – the set of all integers
- $\mathbb{N}$ – the set of all natural numbers
- $\mathbb{Q}$ – the set of all rational numbers (numbers that can be written as the ratio of two integers, $\frac{p}{q}$, where $q \neq 0$)
- $\mathbb{C}$ – the set of all complex numbers (numbers of the form $a + bi$, where a and b are real numbers and $i = \sqrt{-1}$)

A superscripted plus or minus sign is used to restrict the set to positive or negative numbers. for instance, $\mathbb{R}^+$ would be the set of all positive real numbers, and $\mathbb{Z}^-$ would be the set of all negative integers.

Some of these sets are, of course, subsets of others. Specifically, $\mathbb{N} \subset \mathbb{Z} \subset \mathbb{Q} \subset \mathbb{R} \subset \mathbb{C}$.

RECURSIVE AND EXPLICIT FORMULAS

Sets of numbers can have a consistent relationship between the elements of the set. Some of these relationships can be represented with simple recursive or explicit formulas. For example, consider the set of all positive, even numbers (a) and the set of all positive, odd numbers (b):

<u>POSITIVE, EVEN NUMBERS:</u>

Recursive: $a_n = a_{n-1} + 2, n \geq 2, a_1 = 0$
Explicit: $a_n = 2n, n \geq 0$

<u>POSITIVE, ODD NUMBERS:</u>

Recursive: $b_n = b_{n-1} + 2, n \geq 2, b_1 = 1$
Explicit: $b_n = 2n + 1, n \geq 0$

Each set of numbers represents a linear function, with a constant rate of change of 2. The positive, even numbers represent a linear function that is proportional, whereas the positive, odd numbers represent a linear function that is not proportional. The set of even, positive numbers is represented by a function with a y-intercept of 0. The set of odd, positive numbers is represented by a function with a y-intercept of 1.

Venn Diagrams

Set Operations with Venn Diagrams

A Venn diagram is a useful visual tool for representing two or three sets and their common elements. Each set is drawn as a **circle**, with the different circles overlapping. Elements are placed in the circle corresponding to the appropriate set or sets—or placed outside all the circles if they belong to the universe of discourse but not to any of the sets.

For example, suppose our universe of discourse is the integers from 1 to 9, and we have the three sets $A = \{1, 2, 3, 4, 5, 6\}$, $B = \{4, 5, 6, 7\}$, and $C = \{3, 6, 9\}$. This could be illustrated with the following **Venn diagram**:

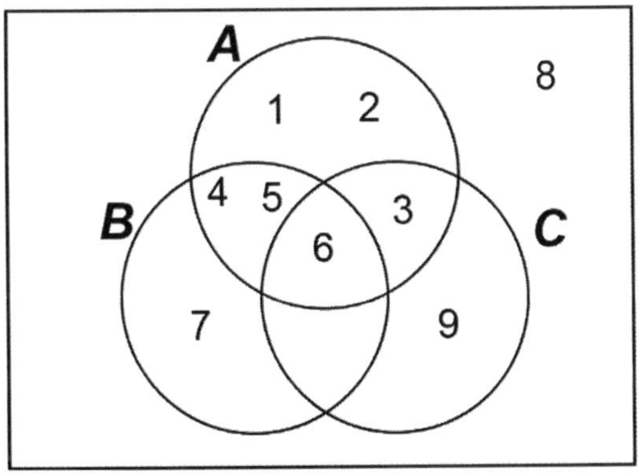

Note, for example, that 6 appears in the center of the diagram where all three circles overlap, because it is an element of all three sets. On the other hand, 8 is placed outside the three circles because it is not an element of any of the sets. Sometimes, instead of writing the elements themselves in the diagram, the total *number* of elements in each part of the Venn diagram is noted. This is especially useful for solving problems involving these numbers of elements.

Using Venn Diagrams to Solve Problems

Venn diagrams are useful for solving problems involving the numbers of elements in sets and in their intersections. By putting those numbers into a Venn diagram, it's simple to see how many must be in the "leftover" parts.

For example, suppose we're told that 200 voters were polled about two propositions, Proposition 1 and Proposition 2. Further,120 support Proposition 1, 85 support Proposition 2, and 50 support both propositions. To find how many of the voters support neither proposition, we can draw a Venn diagram with a circle representing the supporters of each proposition. We know 50 voters support both propositions, so we can write a 50 in the center of the diagram.

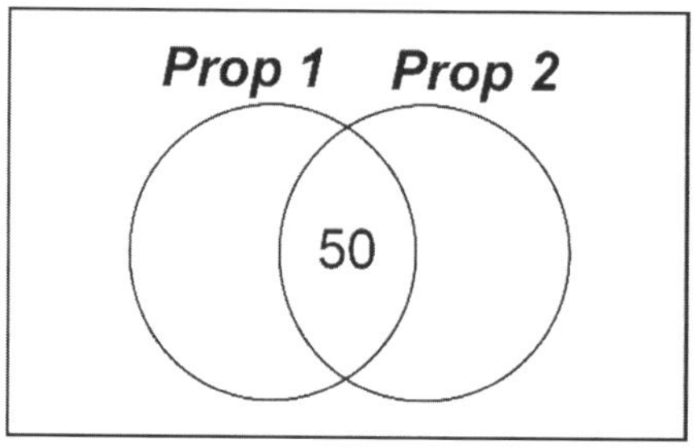

The Proposition 1 circle should contain 120 voters total, so subtracting the 50 voters in the overlap, the other section of the circle must contain $120 - 50 = 70$. Similarly, the nonoverlapping part of the Proposition 2 circle should contain $85 - 50 = 35$.

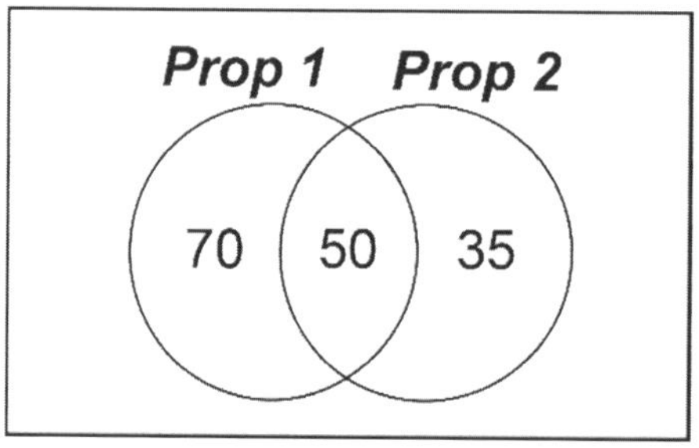

Adding all three sections, the total number of voters supporting either proposition is $70 + 50 + 35 = 155$, so there must be $200 - 155 = 45$ voters who support neither.

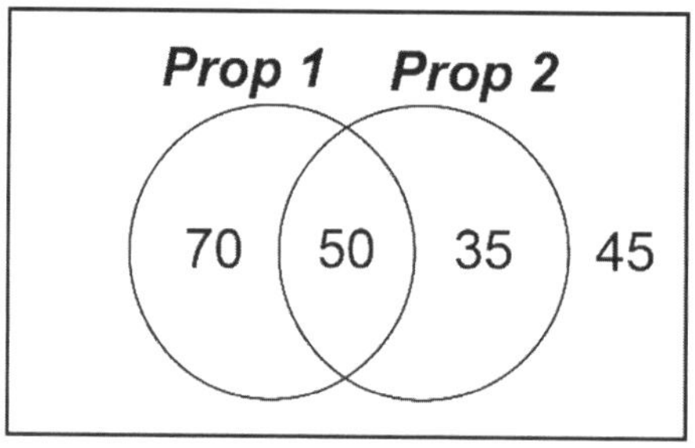

Advanced Set Theory

Properties of Infinite Sets

An infinite set has an infinite number of elements. a more technical definition is that a set is **infinite** if its elements can be put in one-to-one correspondence with the elements of one of its proper subsets. For instance, consider the set of all positive integers, $\mathbb{Z}^+$. If we remove the number 1, we can still match up every positive integer to an element of the remaining subset (by simply matching the number n to the number $n + 1$). $\mathbb{Z}^+$ is therefore an infinite set.

The union of an infinite set and any other set is always infinite. The intersection of two infinite sets may or may not be infinite. For instance, consider the set P of prime numbers, the set O of positive odd integers, and the set E of positive even integers—all infinite sets. $P \cap O$ is another infinite set, because there are infinitely many odd prime numbers. $P \cap E$ is finite, because there is only one even prime number. $E \cap O$ is the empty set. Likewise, the complement of an infinite set may be infinite, finite, or the empty set (if the set equals the universe of discourse).

Cardinality

Although all infinite sets have infinitely many members, they may still have different sizes. Two sets are defined to have the same size—or more technically the same **cardinality**—if each element of one set can be matched up to a unique element of the other, with none left over. For example, the set of all integers, $\mathbb{Z}$, and the set of all *even* integers have the same cardinality, even though the latter is a proper subset of the former: every number n in $\mathbb{Z}$ can be matched to a unique number $2n$ in the set of all *even* integers. Although it's more difficult to prove, the set of all rational numbers, $\mathbb{Q}$, also has the same cardinality as $\mathbb{Z}$. However, the set of all real numbers, $\mathbb{R}$, does *not* have the same cardinality—it is impossible to match each integer to a unique real number without any real numbers left over. The most famous proof of this fact was developed by Georg Cantor, and is called the *diagonal argument.*

Sets with the same cardinality as $\mathbb{Z}$ are said to be "countably infinite," or simply **countable**. Sets with a larger cardinality are said to be "uncountably infinite," or **uncountable**.

Cartesian Products and Binary Relations

A **Cartesian product** is the product of two sets of data, X and Y, such that all elements x are a member of set X, and all elements y are a member of set Y. The product of the two sets, $X \times Y$ is the set of all ordered pairs (x, y). For example, given a standard deck of 52 playing cards, there are four possible suits (hearts, diamonds, clubs, and spades) and thirteen possible card values (the numbers 2 through 10, ace, jack, queen, and king). If the card suits are set X and the card values are set Y, then there are $4 \times 13 = 52$ possible different (x, y) combinations, as seen in the 52 cards of a standard deck.

A **binary relation**, also referred to as a relation, dyadic relation, or 2-place relation, is a subset of a Cartesian product. It shows the relation between one set of objects and a second set of objects, or between one set of objects and itself. The prefix *bi-* means *two*, so there are always two sets involved—either two different sets, or the same set used twice. The ordered pairs of the Cartesian product are used to indicate a binary relation. Relations are possible for situations involving more than two sets, but those are not called binary relations.

The five types of relations are reflexive, symmetric, transitive, antisymmetric, and equivalence. A **reflexive relation** has $x\Re x$ (x related to x) for all values of x in the set. A **symmetric relation** has $x\Re y \Rightarrow y\Re x$ for all values of x and y in the set. A **transitive relation** has ($x\Re y$ and $y\Re z$) $\Rightarrow x\Re z$ for all values of x, y, and z in the set. An **antisymmetric relation** has ($x\Re y$ and $y\Re x$) $\Rightarrow x = y$ for all values of x and y in the set. A relation that is reflexive, symmetric, and transitive is called an **equivalence relation**.

VERTEX-EDGE GRAPHS

A **vertex-edge graph** is a set of items or objects connected by pathways or links. As an example, consider the following set of nodes and a few graphs representing ways of connecting them with edges.

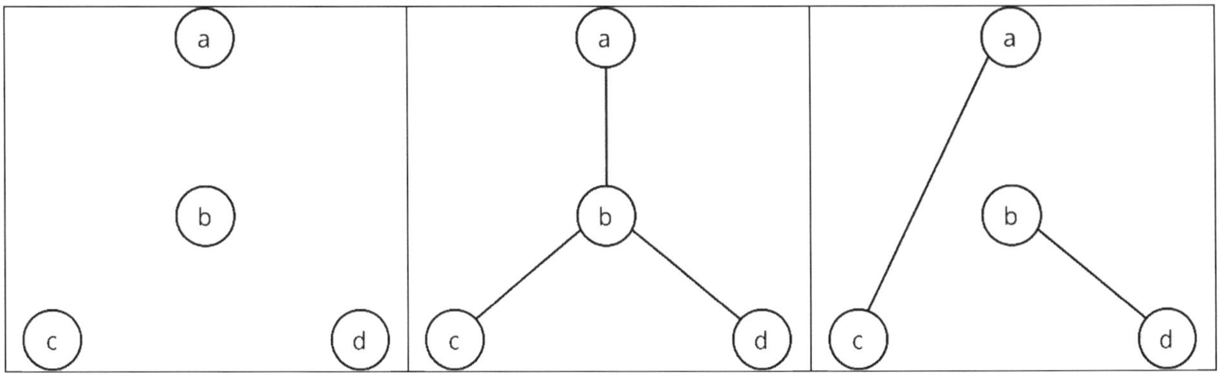

Vertex-edge graphs are useful for solving problems involving schedules, relationships, networks, or paths among a set number of objects. The number of objects may be large, but it will never be infinite. The **vertices** or points on the graph represent the objects and may also be referred to as **nodes**. The nodes are joined by line segments called **edges** or links that show the specific paths that connect the various elements represented by the nodes. The number of nodes does not have to equal the number of edges. There may be more or less, depending on the number of allowable paths.

An **endpoint** on a vertex-edge graph is a vertex on exactly one edge. In the case of a vertex that is an endpoint, the edge that the vertex is on is incident with the vertex. Two edges are considered to be adjacent if they share a vertex. Two vertices are considered to be adjacent if they share an edge.

In a vertex-edge graph, a **loop** is an edge that has the same vertex as both endpoints. To calculate the **degree of a vertex** in a vertex-edge graph, count the number of edges that are incident with the vertex, counting loops twice since they meet the vertex at both ends. The **degree sum formula** states that the sum of the degrees of all vertices on a vertex-edge graph is always equal to twice the number of edges on the graph. Thus, the sum of the degrees will never be odd, even if there are an odd number of vertices. Consider the following graph. Node d is an endpoint, there is a loop on node b, and the degree sum of the graph is $2 \times 5 = 10$.

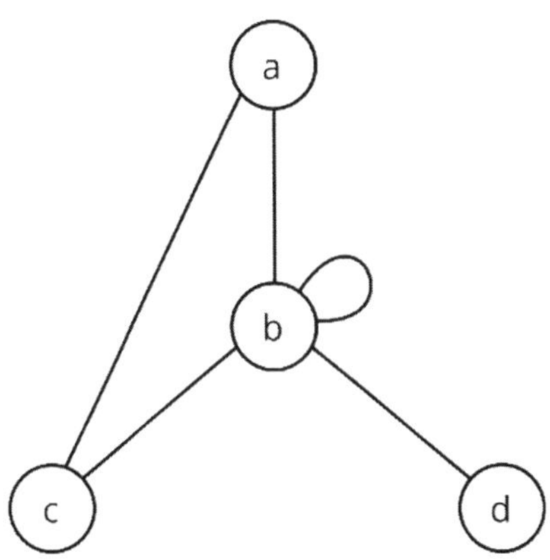

In a vertex-edge graph, a **path** is a given sequence of vertices that follows one or more edges to get from vertex to vertex. There is no jumping over spaces to get from one vertex to the next, although doubling back over an edge already traveled is allowed. A **simple path** is a path that does not repeat an edge in traveling from beginning to end. Think of the vertex-edge graph as a map, with the vertices as cities on the map, and the edges as roads between the cities. To get from one city to another, you must drive on the roads. A simple path allows you to complete your trip without driving on the same road twice.

In a vertex-edge graph, a **circuit** is a path that has the same starting and stopping point. Picturing the vertex-edge graph as a map with cities and roads, a circuit is like leaving home on vacation and then returning home after you have visited your intended destinations. You may go in one direction and then turn around, or you may go in a circle. A **simple circuit** on the graph completes the circuit without repeating an edge. This is like going on vacation and returning home without driving on the same road twice.

Graph

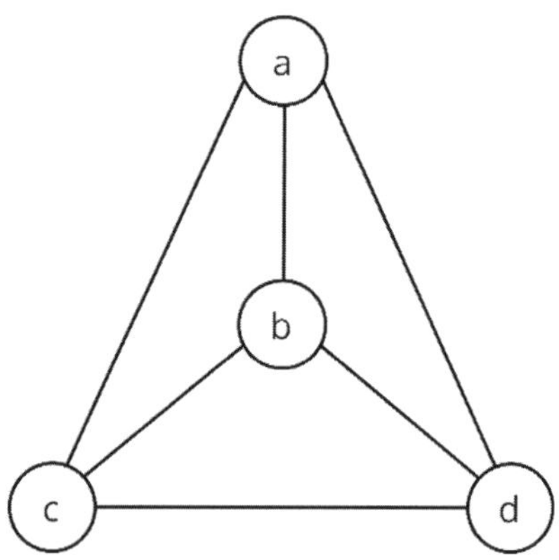

Example Paths

Path:
a-b-c-b-d

Simple path:
a-b-d

Circuit:
b-a-c-a-b

Simple circuit:
a-b-c-a

On a vertex-edge graph, any path that uses each edge exactly one time is called an **Euler path**. One simple way to rule out the possibility of an Euler path is to calculate the degree of each vertex. If more than two vertices have an odd degree, an Euler path is impossible. A path that uses each vertex exactly one time is called a **Hamiltonian path**.

Graph

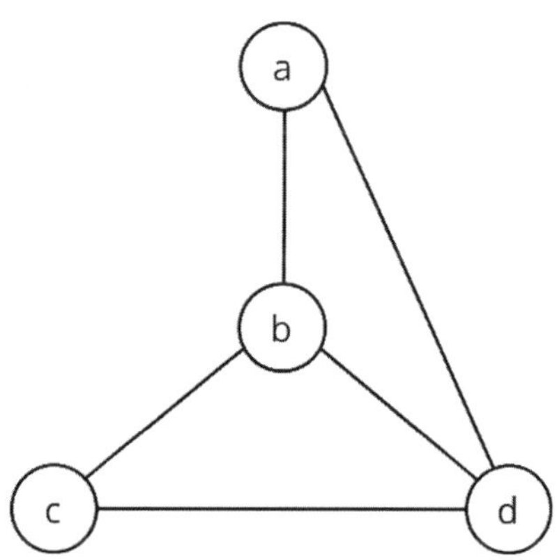

Example Paths

Euler path:
b-c-d-a-b-d

Hamiltonian path:
a-b-c-d

If every pair of vertices is joined by an edge, the vertex-edge graph is said to be **complete**. If the vertex-edge graph has no simple circuits in it, then the graph is said to be a **tree**. If every vertex is connected to every other vertex by some *path*, then the graph is said to be **connected**, otherwise it is **disconnected**.

Complete and Connected

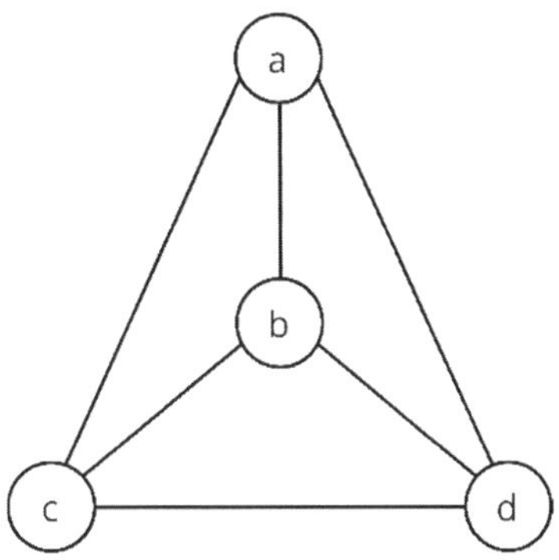

Tree and Connected

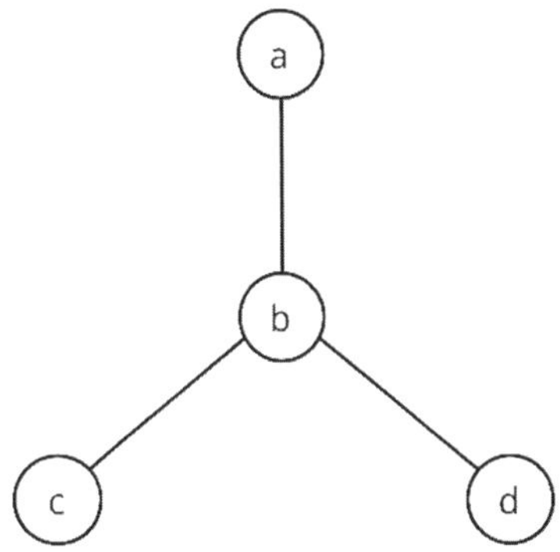

Disconnected

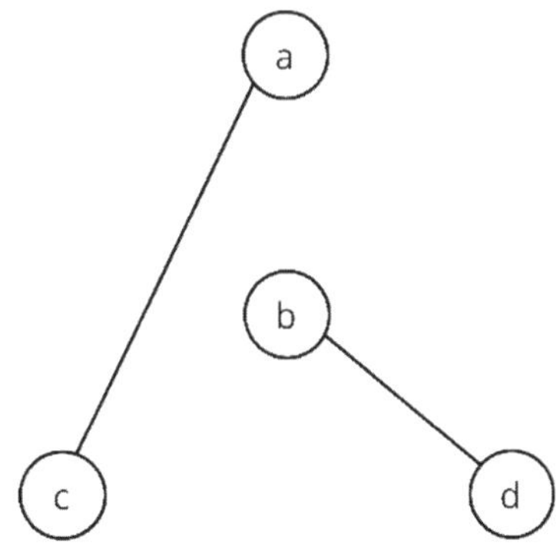

Chapter Quiz

Ready to see how well you retained what you just read? Scan the QR code to go directly to the chapter quiz interface for this study guide. If you're using a computer, simply visit the online resources page at **mometrix.com/resources719/orelamath** and click the Chapter Quizzes link.

ORELA Practice Test

Want to take this practice test in an online interactive format?
Check out the online resources page, which includes interactive practice questions and much more: **mometrix.com/resources719/orelamath**

1. Which of these defines the recursive sequence $a_1 = 2, a_{n+1} = a_n - 3$ explicitly?

a. $a_n = 2n - 3$
b. $a_n = -3n + 5$
c. $a_n = n - 1$
d. $a_n = -2n + 2$

2. Will needed his dinner to include at least 50 grams of protein, so he wanted to make sure he was eating enough chicken. He recalled that chicken has approximately 25 grams of protein in a 4-oz serving. He didn't have a scale that weighed ounces, but he had a 5-lb bag of flour that he planned to divide into equal portions to approximate the weight of chicken he needed, so that he could compare the weight of the flour to the chicken. Which of the following shows the calculation for the number of portions Will must divide the flour into?

a. $50 \times \frac{4}{25} \times \frac{1}{16} \times 5$
b. $25 \times 4 \times \frac{1}{16} \times \frac{1}{5}$
c. $1 \times 5 \times 16 \times \frac{25}{4} \times \frac{1}{50}$
d. $\frac{25}{4} \times \frac{16}{1} \times \frac{1}{5} \times 50$

3. Which number comes next in the sequence: 1, 4, 9, 16, ...?

a. 20
b. 24
c. 25
d. 32

4. A pasta salad was chilled in the refrigerator at 35° F overnight for 9 hours. The temperature of the pasta dish dropped from 86° F to 38° F. What was the average rate of cooling per hour?

a. 4.8°/hr
b. 5.3°/hr
c. 5.15°/hr
d. 0.532°/hr

5. A rainbow pattern is designed from semicircles, as shown below.

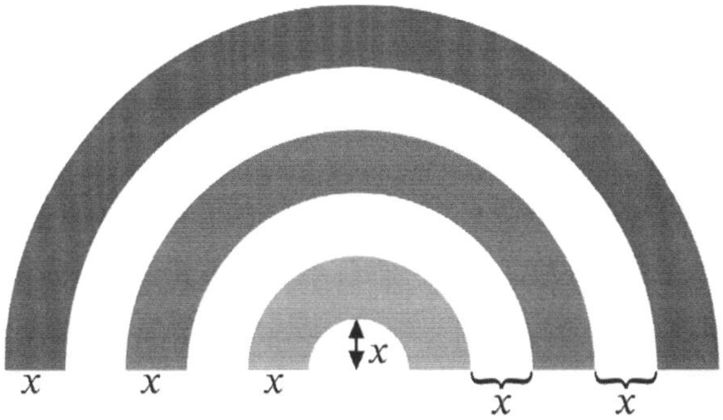

Which of the following gives the area A of the shaded region as a function of x?

a. $A = \frac{21\pi x^2}{2}$
b. $A = 21\pi x^2$
c. $A = 42\pi x^2$
d. $A = 82\pi x^2$

6. A company is designing a paperweight in the shape of a square pyramid. Each of the sides of the square base measures x in, and the height is z in. If the length of each side of the base were increased by $\frac{1}{2}$ in, how much more volume would the paperweight have?

a. $\frac{z\left(x^2+x+\frac{z}{4}\right)}{3}$ in^3
b. $\frac{x^2z+4}{12}$in^3
c. $\frac{z(4x+1)}{12}$ in^3
d. $\frac{1}{4}$in^3

7. Which number should fill in the blank in the sequence 18, 23, 27, ___, 32, 33?

a. 28
b. 29
c. 30
d. 31

8. Two even integers and one odd integer are multiplied together. Which of the following could be their product?

a. 3.75
b. 9
c. 16.2
d. 24

Refer to the following for questions 9–11:

Consider the following graphic showing demographics of a high school with 1,219 total students:

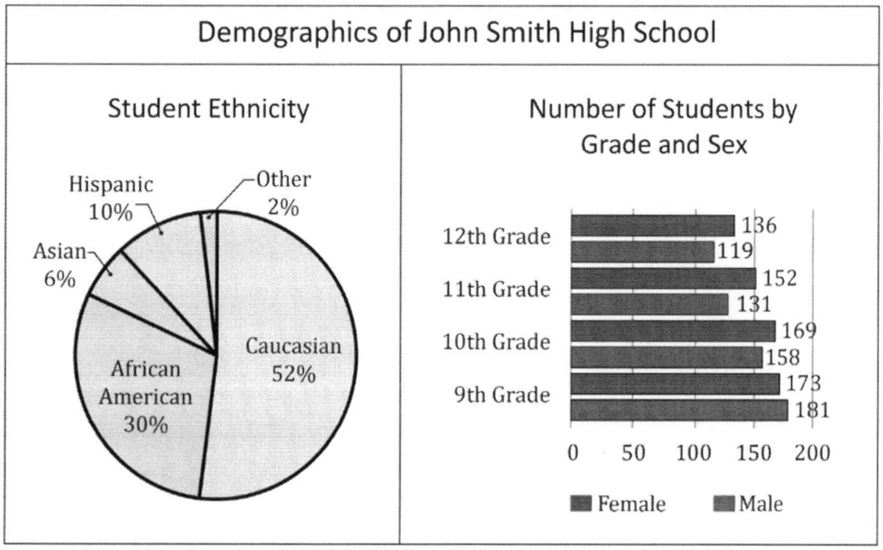

9. Which of these is the greatest quantity?

a. The average number of male students in the 11th and 12th grades
b. The number of Hispanic students at the school
c. The difference in the number of male and female students at the school
d. The difference in the number of 9th and 12th grade students at the school

10. An eleventh grader is chosen at random to represent the school at a conference. What is the approximate probability that the student is male?

a. 0.03
b. 0.11
c. 0.22
d. 0.46

11. Compare the two quantities.

Quantity A	**Quantity B**
The percentage of Caucasian students	The percentage of female students, rounded to the nearest whole number

a. Quantity A is greater.
b. Quantity B is greater.
c. The two quantities are the same.
d. The relationship cannot be determined from the given information.

12. A teacher wishes to divide her class of 20 students into 4 groups, each of which will have 3 boys and 2 girls. How many different ways are there to form a single group?

a. 248
b. 6,160
c. 73,920
d. 95,040

13. If the midpoint of a line segment graphed on the xy-coordinate plane is $(-2, 3)$ and the slope of the line segment is $-\frac{1}{2}$, which of these is a possible endpoint of the line segment?

a. $(-4,3)$
b. $(2,1)$
c. $(-6,6)$
d. $\left(0,\frac{3}{2}\right)$

14. Which of these is the equation of the function graphed below?

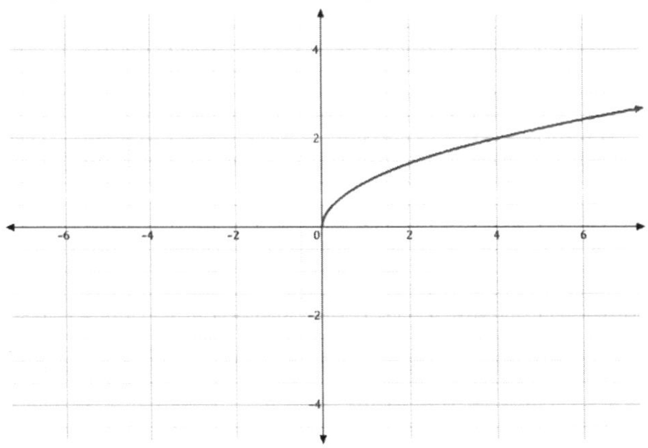

a. $f(x) = x^2$
b. $f(x) = \sqrt{x}$
c. $f(x) = 2^x$
d. $f(x) = \log_2 x$

15. Which of these is a major contribution of the Babylonian civilization to the historical development of mathematics?

a. The division of an hour into 60 minutes, and a minute into 60 seconds, and a circle into 360 degrees
b. The development of algebra as a discipline separate from geometry
c. The use of deductive reasoning in geometric proofs
d. The introduction of Boolean logic

16. Simplify:

$$\begin{bmatrix} 2 & 0 & -5 \end{bmatrix} \left(\begin{bmatrix} 4 \\ 2 \\ -1 \end{bmatrix} - \begin{bmatrix} 3 \\ 5 \\ -5 \end{bmatrix} \right)$$

a. $[-18]$
b. $\begin{bmatrix} 2 \\ 0 \\ -20 \end{bmatrix}$

c. $\begin{bmatrix} 2 & 0 & -20 \end{bmatrix}$

d. $\begin{bmatrix} 2 & 0 & -5 \\ -6 & 0 & 15 \\ 8 & 0 & -20 \end{bmatrix}$

17. Which of these graphs is NOT representative of the data set shown below?

$$\begin{array}{l|l} 3 & 6799 \\ 4 & 23889 \\ 5 & 011158 \\ 6 & 00123 \end{array}$$

Key
$2|123 = 21, 22, 23$

a. 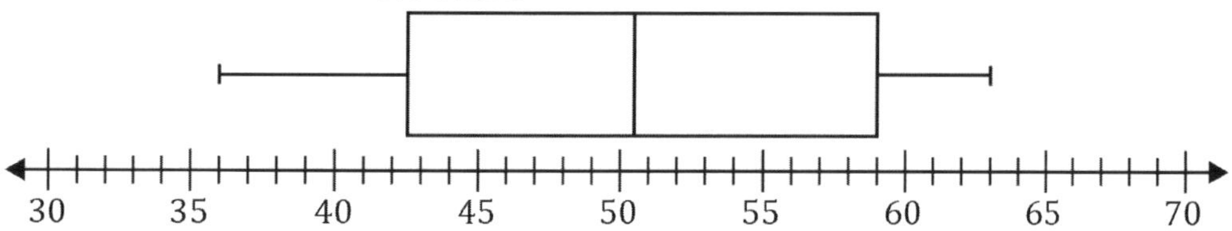

b. 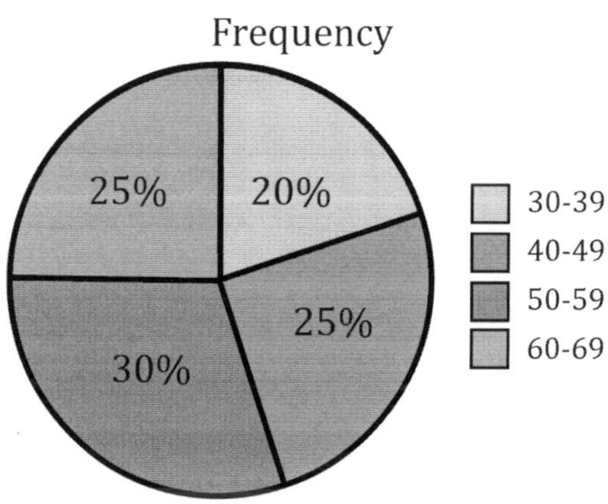

c. 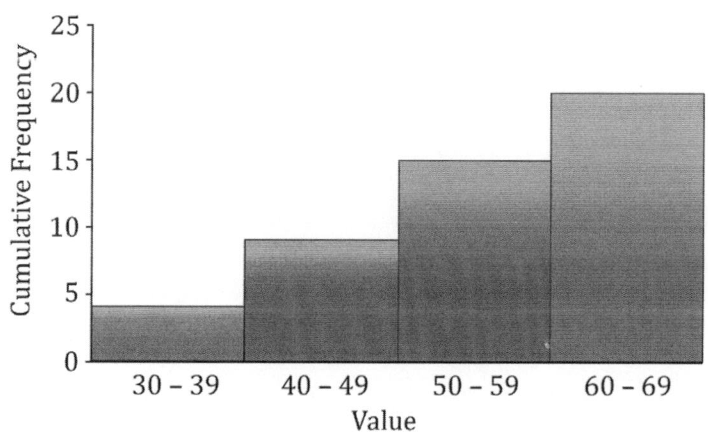

d. All of these graphs represent the data set.

18. Solve $5x^2 - 4x = -2$.

a. $x = \frac{2 \pm \sqrt{14}i}{5}$
b. $x = \pm i\sqrt{24}$
c. $x = \pm\frac{2\sqrt{2}i}{5}$
d. $x = \frac{2 \pm i\sqrt{6}}{5}$

19. Which of the following statements is (are) true when $f(x) = \frac{x^2 - x - 6}{x^3 + 2x^2 - x - 2}$?

I. The graph $f(x)$ has vertical asymptotes at $x = -2$, $x = -1$, and $x = 1$.
II. The x- and y-intercepts of the graph of $f(x)$ are both 3.

a. I only
b. II only
c. I and II
d. Neither statement is true.

20. A shirt is marked down by 25% and placed on a clearance rack, on which is posted a sign reading, "Take an extra 30% off already reduced merchandise." What fraction of the original price is the final sale price of the shirt?

a. $\frac{9}{20}$
b. $\frac{11}{20}$
c. $\frac{21}{40}$
d. $\frac{7}{10}$

21. Identify a possible cross-section polygon NOT formed by a plane containing the given points on the cube.

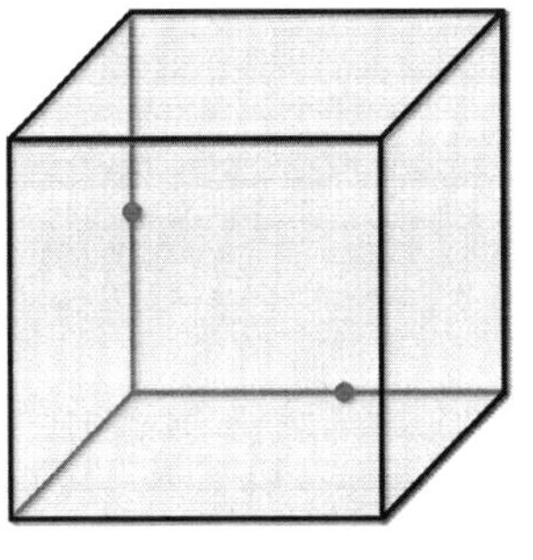

a. Triangle
b. Square
c. Pentagon
d. Hexagon

22. For the holidays, Luke's Bakery offered three seasonal items: pumpkin bread, cranberry muffins, and their famous eggnog pie. They received 57 orders for pumpkin bread, 37 orders for cranberry muffins, and 70 orders for eggnog pie. Fourteen customers ordered both pumpkin bread and cranberry muffins, ten customers ordered both cranberry muffins and eggnog pie, and fifteen customers ordered both pumpkin bread and eggnog pie. Six customers ordered all three items. How many customers placed an order for seasonal items?

a. 112
b. 131
c. 149
d. 164

23. Which of these statements is (are) true for the function $g(x)$?

$$g(x) = \begin{cases} -x+3 & x<2 \\ 2x-1 & x \geq 2 \end{cases}$$

I. $g(3) = 0$
II. The graph of $g(x)$ is discontinuous at $x = 2$.
III. The range of $g(x)$ is all real numbers.

a. II only
b. III only
c. I and II
d. II and III

Refer to the following for questions 24–25:

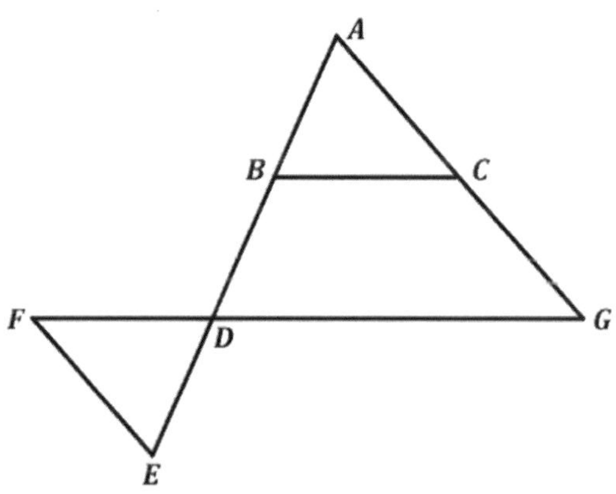

Statement	Reason
1. $\overline{BC} \parallel \overline{FG}$	Given
2.	
3. $\overline{FD} \cong \overline{BC}$	Given
4. $\overline{AB} \cong \overline{DE}$	Given
5. $\Delta ABC \cong \Delta EDF$	________
6. ________	
7. $\overline{FE} \parallel \overline{AG}$	

Given: $\overline{BC} \parallel \overline{FG}; \overline{FD} \cong \overline{BC}; \overline{AB} \cong \overline{DE}$

Prove: $\overline{FE} \parallel \overline{AG}$

24. Which of the following justifies step 5 in the proof?

a. AAS
b. SSS
c. ASA
d. SAS

25. Step 6 in the proof should contain which of the following statements?

a. $\angle BAC \cong \angle DEF$
b. $\angle ABC \cong \angle EDF$
c. $\angle ACB \cong \angle EFD$
d. $\angle GDA \cong \angle EDF$

26. Which of these is a net of a triangular pyramid?

a. 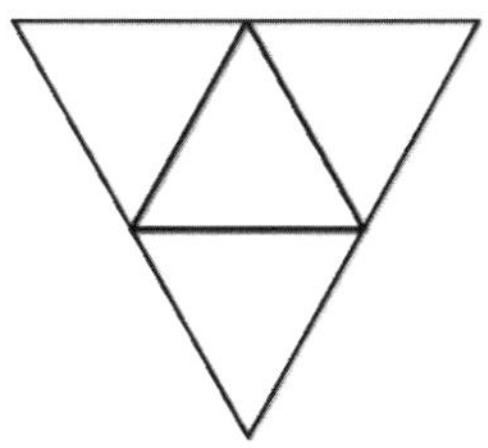

b. 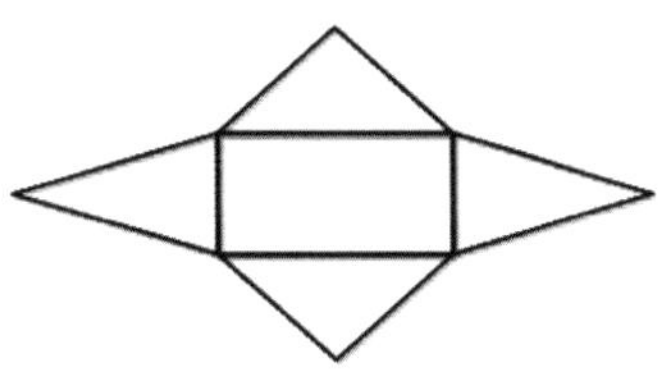

c. 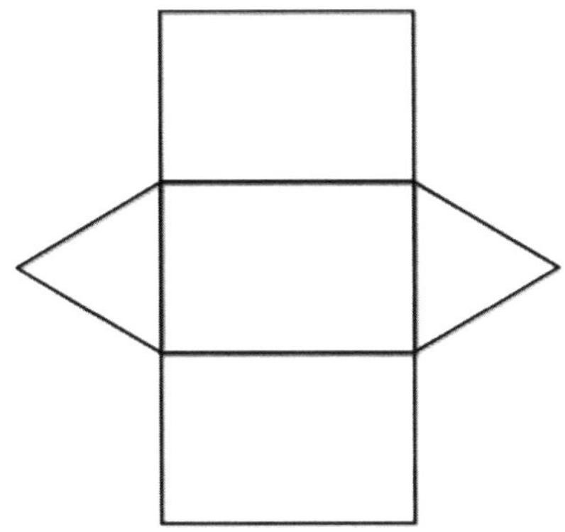

d. 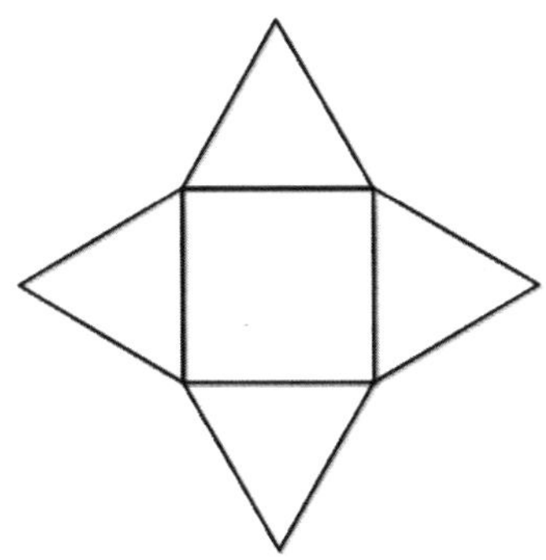

27. What is the exact value of $\tan\left(-\frac{\pi}{4}\right)$?

a. -1
b. $-\frac{\sqrt{2}}{2}$
c. $\frac{\sqrt{2}}{2}$
d. 1

28. In the triangle, which of the following is equal to $\sin\theta$?

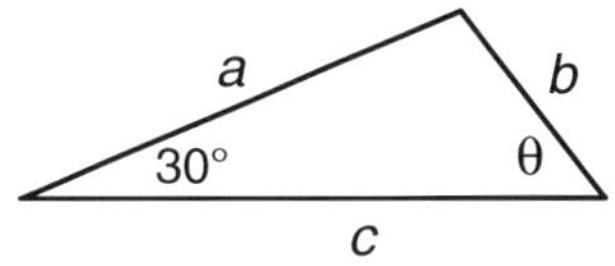

a. $\frac{a}{2b}$
b. $\frac{2b}{a}$
c. $\frac{a}{2c}$
d. $\frac{b}{ac}$

29. In order to analyze the real estate market for two different zip codes within the city, a realtor examined the most recent 100 home sales in each zip code. She considered a house which sold within the first month of its listing to have a market time of one month; likewise, she considered a house to have a market time of two months if it sold after having been on the market for more than one month but before the end of the second month. Using this definition of market time, she determined the frequency of sales by the number of months on the market. The results are displayed below. Which of the following is a true statement for these data?

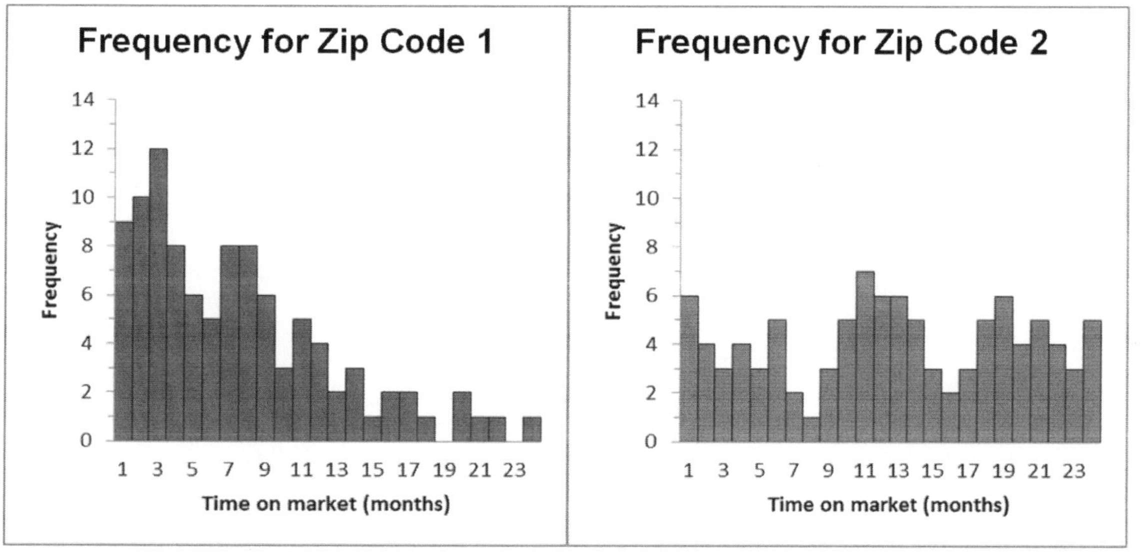

a. The median time a house spent on the market in Zip Code 1 is five months less than Zip Code 2.
b. On average, a house spent seven months longer on the market in Zip Code 2 than in Zip Code 1.
c. The mode time on the market is higher for Zip Code 1 than for Zip Code 2.
d. The median time on the market is less than the mean time on the market for Zip Code 1.

30. If the square of twice the sum of x and three is equal to the product of twenty-four and x, which of these is a possible value of x?

a. $6 + 3\sqrt{2}$
b. $\frac{3}{2}$
c. $-3i$
d. -3

31. The ratio of Jon's monthly expenses to his savings is 7: 1. If he is able to save $725 per month, what are his monthly expenses?

a. $5,075
b. $5,800
c. $6,125
d. $6,350

32. A car is driving along the highway at a constant speed when it runs over a pebble, which becomes lodged in one of the tire's treads. If this graph represents the height h of the pebble above the road in inches (y-axis) as a function of time t in seconds (x-axis), which of these statements is true?

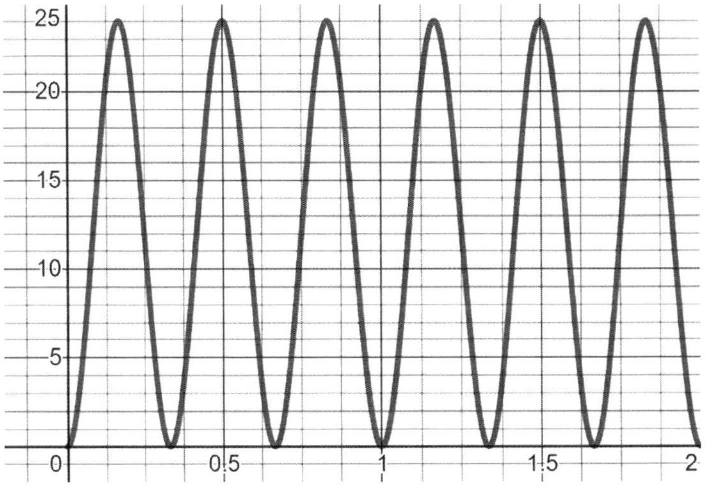

a. The outer diameter of the tire is 12 inches, and the tire rotates 60 times per minute.
b. The outer diameter of the tire is 24 inches, and the tire rotates 60 times per minute.
c. The outer diameter of the tire is 12 inches, and the tire rotates 180 times per minute.
d. The outer diameter of the tire is 24 inches, and the tire rotates 180 times per minute.

33. Kayla has a $75 budget to purchase gifts for her colleagues. She wants to buy coffee mugs and note pads. She may purchase a maximum of 30 items. Each coffee mug costs $6 and each note pad costs $3. Which of the following graphs correctly shows the possible combinations of coffee mugs and note pads that she may buy?

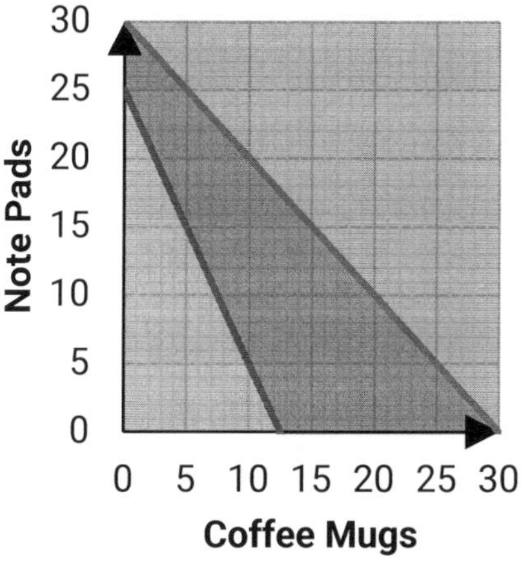

a.

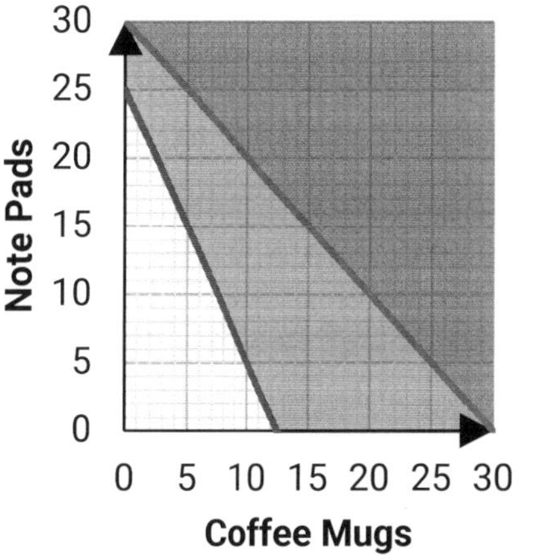

b.

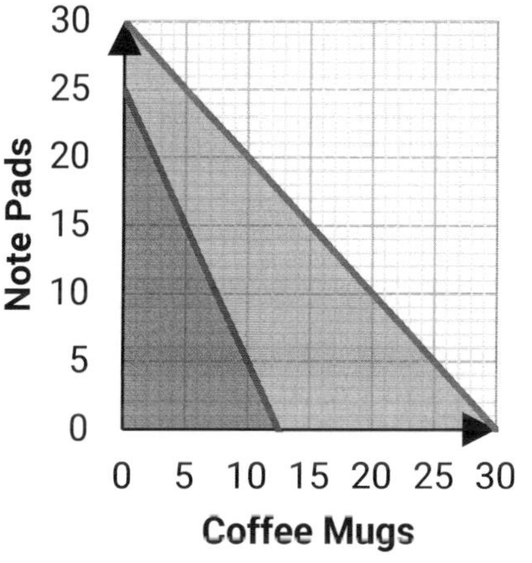

c.

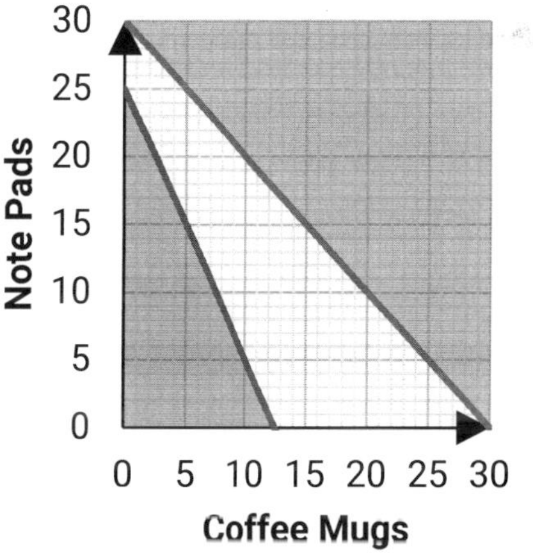

d.

Refer to the following for question 34:

Below are graphed functions $f(x) = a_1 \sin(b_1 x)$ and $g(x) = a_2 \cos(b_2 x)$; a_1 and a_2 are integers, and b_1 and b_2 are positive rational numbers.

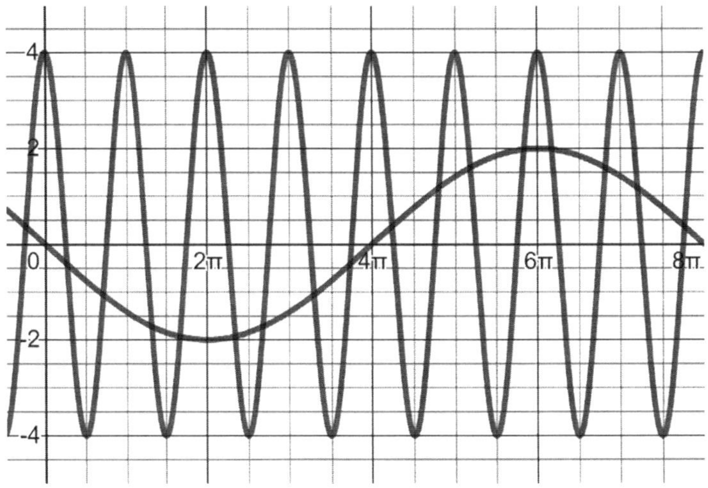

34. Which of the following statements is true?

a. $0 < a_2 < a_1$
b. $a_2 < 0 < a_1$
c. $0 < a_1 < a_2$
d. $a_1 < 0 < a_2$

35. A teacher sits on the floor with five of her students, seated in a circle. If she must sit next to Andrew, how many distinguishable seating arrangements are there? (If an arrangement is merely a rotation of another arrangement, the two are not distinguishable.)

a. 12
b. 24
c. 48
d. 96

36. How many different seven-digit telephone numbers can be created in which no digit repeats and in which zero cannot be the first digit?

a. 5,040
b. 35,280
c. 544,320
d. 3,265,920

37. A circle is inscribed inside quadrilateral $ABCD$. $\overline{CD}$ is bisected by the point at which it is tangent to the circle. If $\overline{AB} = 10$, $\overline{BC} = 12$, and $\overline{CD} = 8$, then what is the length of $\overline{AD}$?

a. $\overline{AD} = 11$
b. $\overline{AD} = 7.5$
c. $\overline{AD} = 6$
d. $\overline{AD} = 2\sqrt{11}$

38. A music appreciation class includes 10 six-year-olds, 9 seven-year-olds, 13 eight-year-olds, 10 nine-year-olds, and 14 ten-year-olds. If a student is randomly selected to demonstrate an instrument, what is the probability that he or she is at least nine years old?

a. $\frac{37}{56}$
b. $\frac{22}{63}$
c. $\frac{4}{7}$
d. $\frac{3}{7}$

39. Which of the graphs shown represents $f(x) = -2|-x+4| - 1$?

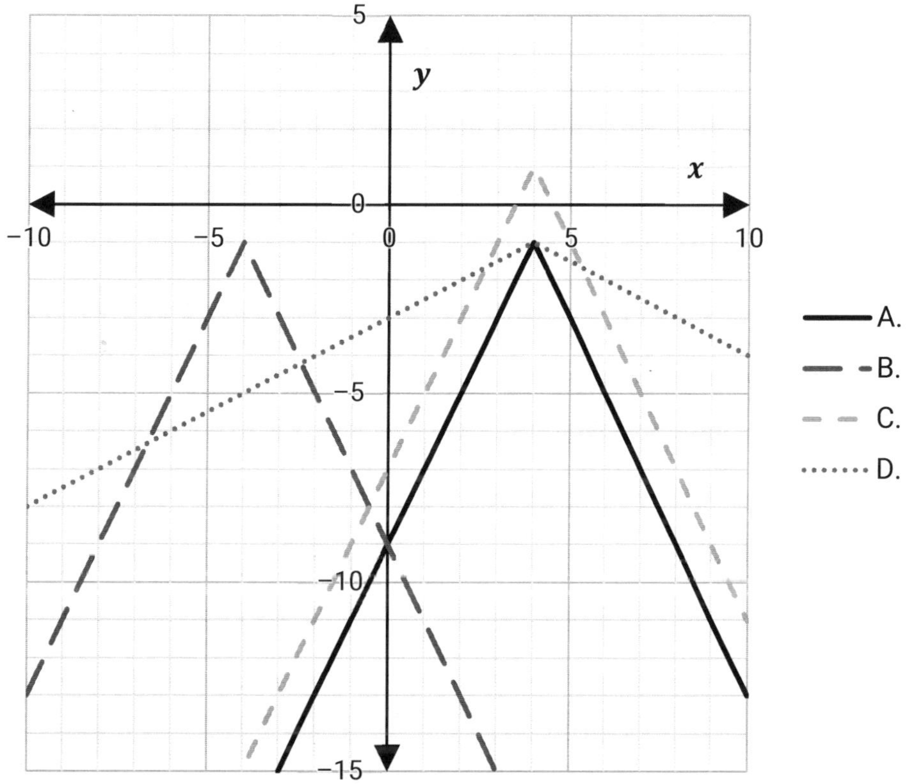

a.
b. 
c. 
d. 

40. The cost of admission to a theme park is shown below.

Under age 10	Ages 10–65	Over age 65
\$15	\$25	\$20

Yesterday, the theme park sold 810 tickets and earned \$14,500. There were twice as many children under 10 at the park as there were other visitors. If x, y, and z represent the number of \$15, \$25, and \$20 tickets sold, respectively, which of the following can be used to find how many of each sold?

a. $\begin{bmatrix} 1 & 1 & 1 \\ 15 & 25 & 20 \\ 1 & -2 & -2 \end{bmatrix} \begin{bmatrix} x \\ y \\ z \end{bmatrix} = \begin{bmatrix} 810 \\ 14{,}500 \\ 0 \end{bmatrix}$

b. $\begin{bmatrix} 1 & 1 & 1 \\ 15 & 25 & 20 \\ 1 & -2 & -2 \end{bmatrix} \begin{bmatrix} 810 \\ 14{,}500 \\ 0 \end{bmatrix} = \begin{bmatrix} x \\ y \\ z \end{bmatrix}$

c. $\begin{bmatrix} 1 & 15 & 1 \\ 1 & 25 & -2 \\ 1 & 20 & -2 \end{bmatrix} \begin{bmatrix} x \\ y \\ z \end{bmatrix} = \begin{bmatrix} 810 \\ 14{,}500 \\ 0 \end{bmatrix}$

d. $\begin{bmatrix} 1 & 15 & 1 \\ 1 & 25 & -2 \\ 1 & 20 & -2 \end{bmatrix} \begin{bmatrix} 810 \\ 14{,}500 \\ 0 \end{bmatrix} = \begin{bmatrix} x \\ y \\ z \end{bmatrix}$

41. Which of the following expressions is equivalent to $\frac{2+3i}{4-2i}$?

a. $\frac{1}{10} + \frac{4}{5}i$

b. $\frac{1}{10}$

c. $\frac{7}{6} + \frac{2}{3}i$

d. $\frac{1}{10} + \frac{3}{10}i$

Refer to the following for question 42:

The diagrams below show the number of people in a small town who participate in various types of volunteer work.

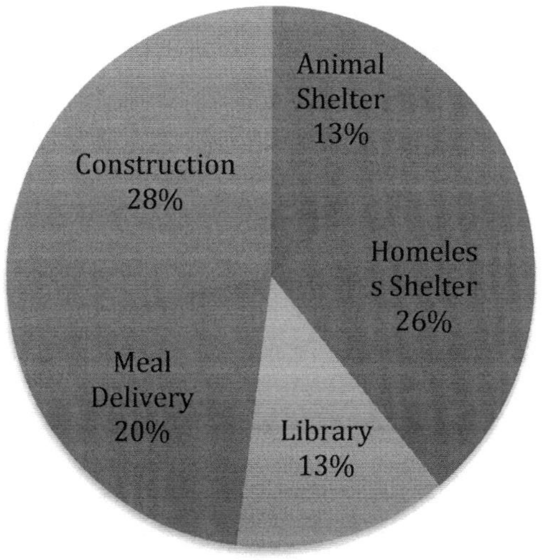

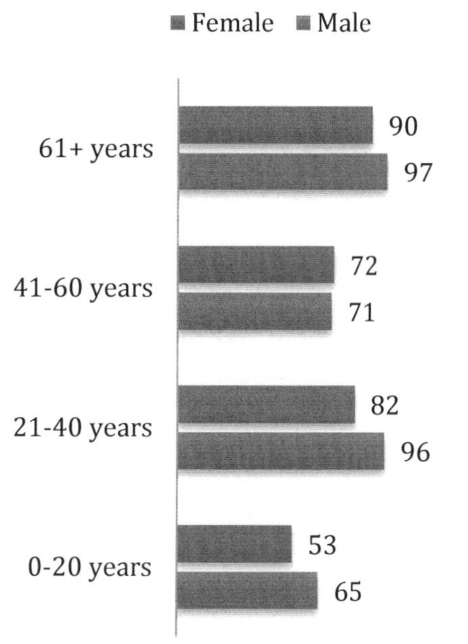

42. Compare the two quantities.

Quantity A	**Quantity B**
The percentage of volunteers in construction	The percentage of volunteers who are females over age 40

a. Quantity A is greater.
b. Quantity B is greater.
c. The two quantities are the same.
d. The relationship cannot be determined from the given information.

43. Which of the following is the best estimate for $23.97124 \div 8.023$?

a. 2
b. 3
c. 16
d. 20

44. Which of these is the equation graphed below?

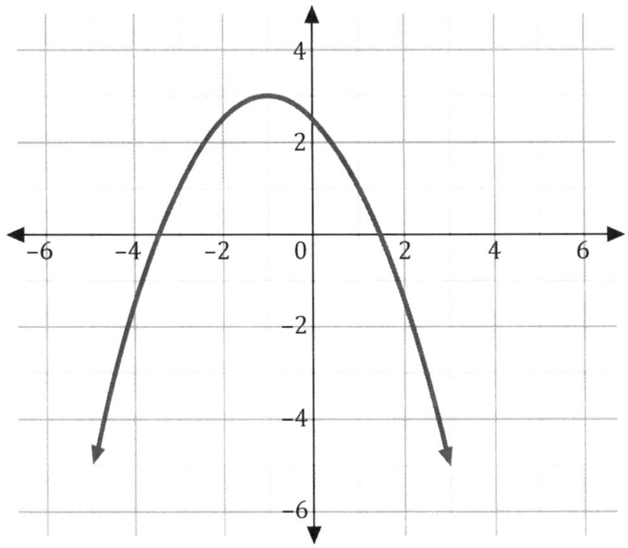

a. $y = -2x^2 - 4x + 1$
b. $y = -x^2 - 2x + 5$
c. $y = -x^2 - 2x + 2$
d. $y = -\frac{1}{2}x^2 - x + \frac{5}{2}$

Refer to the following for questions 45–46:

The box-and-whisker plot displays student test scores assessed throughout a semester to see if students were improving.

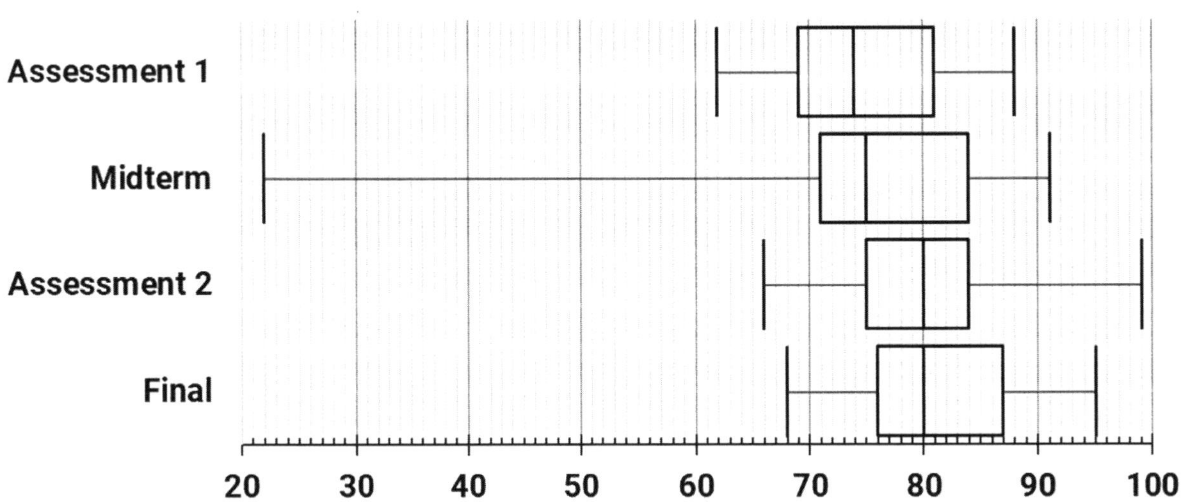

45. Which assessment has the greatest range of test scores?

a. Assessment 1
b. Midterm
c. Assessment 2
d. Final

46. What is the probability that a test score chosen at random from the Final is less than 87?

a. 0.25
b. 0.5
c. 0.6
d. 0.75

47. Given: $f(x) = 10^x$. If $f(x) = 5$, which of these approximates x?

a. 100,000
b. 0.00001
c. 0.7
d. 1.6

48. Which mathematician is responsible for what is often called the most remarkable and beautiful mathematical formula, $e^{i\pi} + 1 = 0$?

a. Pythagoras
b. Euclid
c. Euler
d. Fermat

49. How much $4.171 + 4.55$ is if the sum is rounded to the nearest tenth?

a. 8.8
b. 8.7
c. 8.72
d. 8

50. Which of these statements is NOT necessarily true when $f(x) = \log_b x$ and $b > 1$?

a. The x-intercept of the graph of $f(x)$ is 1.
b. The graph of $f(x)$ passes through $(b, 1)$.
c. $f(x)$ is undefined when $x < 1$.
d. If $g(x) = b^x$, the graph of $f(x)$ is symmetric to the graph of $g(x)$ with respect to $y = x$.

51. Three stores carry the same brand of furniture. Store I offers a leather couch for $3,200, Store II (an online retailer) sells it for $3,050, and Store III sells it for $3,375. Store I is having a 20% off sale, and Store III is offering a $750 discount. Stores I and III charge 8% sales tax, which is assessed after any discounts or coupons are applied. Store II does not charge tax but has a $50 delivery fee. Which of the following lists the stores in order of lowest to highest price after all discounts, taxes, and fees are applied?

a. Store I, Store II, Store III
b. Store II, Store III, Store I
c. Store I, Store III, Store II
d. Store III, Store I, Store II

52. Which of these would best illustrate the distribution of data across a range?

a. Pie chart
b. Line graph
c. Box-and-whisker plot
d. Venn diagram

53. A developer decides to build a fence around a neighborhood park, which is positioned on a rectangular lot. Rather than fencing along the lot line, he fences x feet from each of the lot's boundaries. By fencing a rectangular space 141 yd^2 smaller than the lot, the developer saves \$432 in fencing materials, which cost \$12 per linear foot. How much does he spend?

a. \$160
b. \$456
c. \$3,168
d. The answer cannot be determined from the given information.

54. Which of these demonstrates the relationship between the sets of prime numbers, real numbers, natural numbers, complex numbers, rational numbers, and integers?

$\mathbb{P}$–Prime; $\mathbb{R}$–Real; $\mathbb{N}$–Natural; $\mathbb{C}$–Complex; $\mathbb{Q}$–Rational; $\mathbb{Z}$–Integer

a. $\mathbb{P} \subseteq \mathbb{Q} \subseteq \mathbb{R} \subseteq \mathbb{Z} \subseteq \mathbb{C} \subseteq \mathbb{N}$
b. $\mathbb{P} \subseteq \mathbb{N} \subseteq \mathbb{Z} \subseteq \mathbb{Q} \subseteq \mathbb{R} \subseteq \mathbb{C}$
c. $\mathbb{C} \subseteq \mathbb{R} \subseteq \mathbb{Q} \subseteq \mathbb{Z} \subseteq \mathbb{N} \subseteq \mathbb{P}$
d. None of these

55. On a floor plan drawn at a scale of 1 : 24, the area of a rectangular room is 25 in^2. What is the actual area of the room?

a. 1,440 in^2
b. 144 in^2
c. 1,200 ft^2
d. 100 ft^2

56. Identical rugs are offered for sale at two local shops and one online retailer, designated Stores A, B, and C, respectively. The rug's regular sales price is \$296 at Store A, \$220 at Store B, and \$198.00 at Store C. Stores A and B collect 8% in sales tax on any after-discount price, while Store C collects no tax but charges a \$35 shipping fee. A buyer has a 30% off coupon for Store A and a \$10 off coupon for Store B. Which of these lists the stores in order of lowest to highest final sales price after all discounts, taxes, and fees are applied?

a. Store A, Store B, Store C
b. Store B, Store C, Store A
c. Store C, Store A, Store B
d. Store C, Store B, Store A

57. Solve $x^4 + 64 = 20x^2$.

a. $x = \{2, 4\}$
b. $x = \{-4, -2, 2, 4\}$
c. $x = \{2i, 4i\}$
d. $x = \{-4i, -2i, 2i, 4i\}$

58. Given that x is a whole number greater than 5 and that the greatest common factor of x and y is greater than 1, which of the following must be true?

Quantity A	Quantity B
y	the least common multiple of x and y

a. Quantity A $>$ Quantity B
b. Quantity A $<$ Quantity B
c. Quantity A $=$ Quantity B
d. Quantity A $\leq$ Quantity B

59. Solve $\csc^2\theta = 2\cot\theta$ for $0 < \theta \leq 2\pi$.

a. $\theta = \frac{\pi}{6}$ or $\frac{7\pi}{6}$
b. $\theta = \frac{\pi}{4}$ or $\frac{5\pi}{4}$
c. $\theta = \frac{3\pi}{4}$ or $\frac{7\pi}{4}$
d. $\theta = \frac{3\pi}{4}$ or $\frac{\pi}{6}$

60. Which of these equations is represented by the graph below?

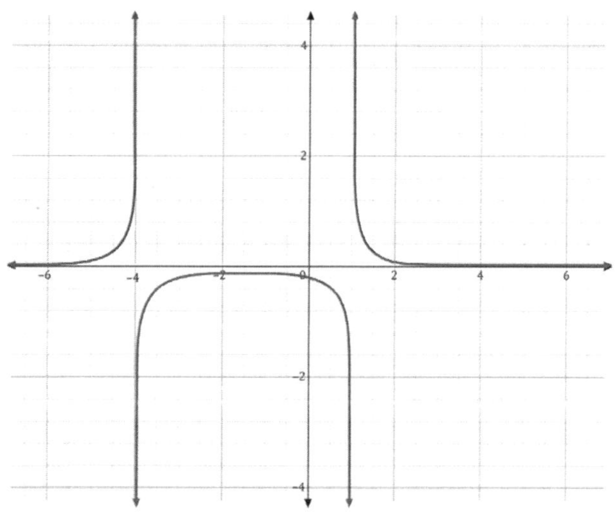

a. $y = \frac{2x-2}{x^2+3x-4}$
b. $y = \frac{1}{x^2+3x-4}$
c. $y = \frac{1}{x+4} + \frac{1}{x-1}$
d. $y = \frac{x}{x^2+3x-4}$

Refer to the following for questions 61–62:

For a science fair project, Paul wanted to find the probability of rolling an even number with a weighted die versus a fair die. He did fifty tests, rolling the die three times for each test and recording the number of even numbers that came up. The chart below shows his results.

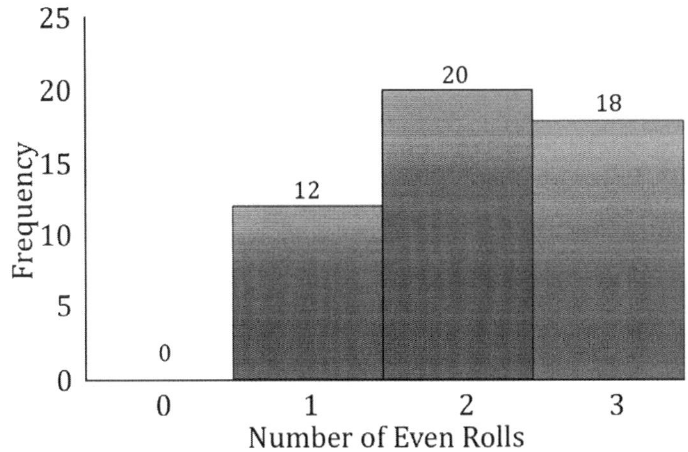

61. Given these experimental data, which of these approximates $P(\text{even})$ for a single roll of this die?

a. 0.468
b. 0.5
c. 0.625
d. 0.707

62. Which of these shows the graphs of the probability distributions (probability of rolling an even) from five rolls of this unevenly-weighted die and five rolls of a fair die?

a.
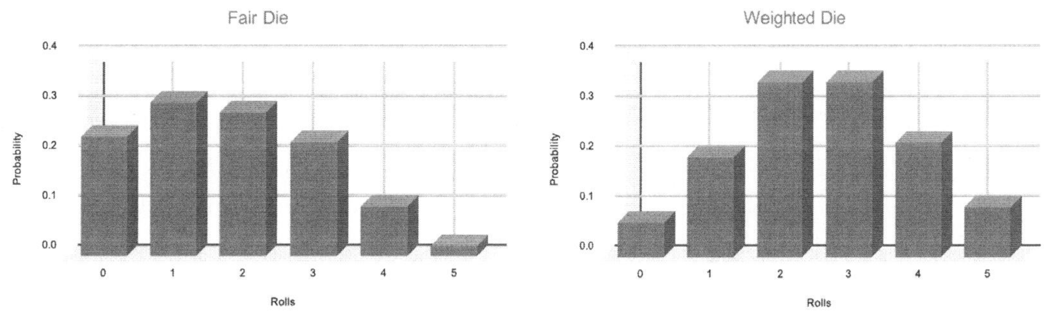

b.
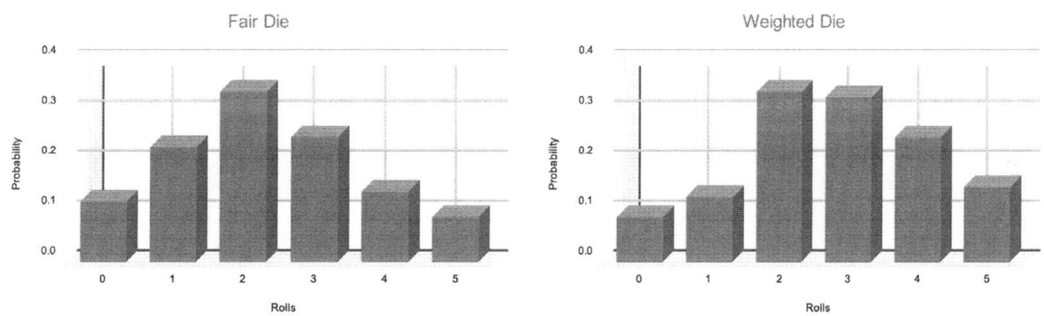

c.
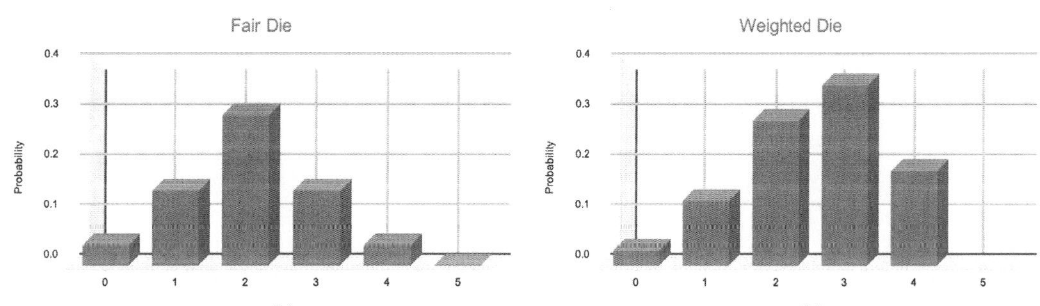

d. None of the above

63. Simplify $(2 - 3i)^2 - (1 - 4i)$.

a. $\sqrt{61}$
b. $-6 - 8i$
c. $6 + 8i$
d. 10

64. A water tower is in the shape of a cylinder with a diameter of 840 in and a height of 220 in. How much water can it hold when completely full?

a. $5.54x10^5$ in^3
b. $1.69x10^6$ in^3
c. $1.22x10^8$ in^3
d. $4.87x10^8$ in^3

65. A rancher is putting up a rectangular cattle pen. His original plan was too expensive, so he changed to a smaller size, subtracting x feet from each of the original boundaries. By fencing a rectangular space 764 yd^2 smaller than the original plan, the rancher saves \$648 in fencing materials, which cost \$9 per linear foot. How much does he spend?

a. \$5,832
b. \$6,552
c. \$6,876
d. \$7,200

66. If $\sin\theta = \frac{\sqrt{3}}{2}$ when $\frac{\pi}{2} < \theta < \pi$, what is the value of θ?

a. $\frac{\pi}{6}$
b. $\frac{3\pi}{4}$
c. $\frac{2\pi}{3}$
d. $\frac{5\pi}{6}$

67. Which of the following best represents the value of x, shown in the triangle below?

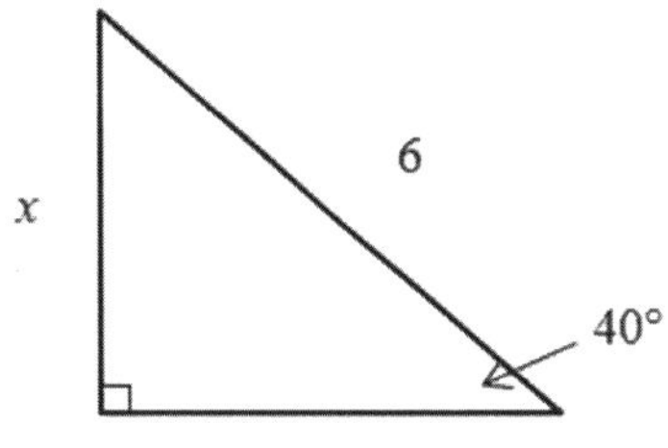

a. 2.6
b. 3.1
c. 3.9
d. 4.4

68. Which of these is equivalent to $\log_y 64$ if $2\log_8 y + \log_8 4 = 2$?

a. 1
b. 2
c. 3
d. 6

69. Which of the following statements is (are) true when $f(x) = \frac{x^2-2x-3}{x^3+3x^2-6x-8}$?

I. The graph $f(x)$ has two vertical asymptotes at $x = 2$ and $x = -4$.
II. The x- and y-intercepts of the graph of $f(x)$ are both 3.

a. I
b. II
c. I and II
d. Neither statement is true.

70. How many diagonals are in a dodecagon?

a. 12
b. 24
c. 54
d. 108

71. A group of 9 children is divided into 3 teams for a game. Each team must have 2 girls and 1 boy. How many possible teams are there?

a. 30
b. 45
c. 72
d. 120

72. The vertices of a polygon are $(2,3)$, $(8,1)$, $(6,-5)$, and $(0,-3)$. Which of the following describes the polygon most specifically?

a. Parallelogram
b. Rhombus
c. Rectangle
d. Square

73. The graph of $f(x)$ is a parabola with a focus of (a,b) and a directrix of $y=-b$, and $g(x)$ represents a transformation of $f(x)$. If the vertex of the graph of $g(x)$ is $(a,0)$, which of these is a possible equation for $g(x)$ for nonzero integers a and b?

a. $g(x) = f(x) + b$
b. $g(x) = -f(x)$
c. $g(x) = f(x + a)$
d. $g(x) = f(x - a) + b$

74. What is the radius of the circle defined by the equation $x^2 + y^2 + 2x - 10y + 18 = 0$?

a. $2\sqrt{2}$
b. 4
c. $2\sqrt{5}$
d. $3\sqrt{2}$

75. Complete the analogy: $\frac{1}{x}$ is to $\frac{1}{y}$ as...

a. x^2 is to y^3
b. e^x is to $\log y, y > 0$
c. x is to $\sqrt{y}, x, y \neq 0$
d. x^2 is to $\sqrt{y}, y > 0$

76. To which of the following sets of numbers does $\frac{1}{4}$ belong?

a. The set of real numbers
b. The set of irrational numbers
c. The set of integers
d. The set of whole numbers

77. Simplify: $\begin{bmatrix} 1 & 4 & -\frac{1}{2} \end{bmatrix} \left(\begin{bmatrix} 2 \\ 0 \\ 4 \end{bmatrix} - \begin{bmatrix} -2 \\ 5 \\ 2 \end{bmatrix} \right)$.

a. $\begin{bmatrix} 4 \\ -20 \\ -1 \end{bmatrix}$

b. $[-17]$

c. $[4 \quad -20 \quad -1]$

d. $\begin{bmatrix} 1 & 0 & 4 \\ 4 & 0 & -5 \\ -\frac{1}{2} & 0 & 2 \end{bmatrix}$

78. Express the area of the given right triangle as a function of x.

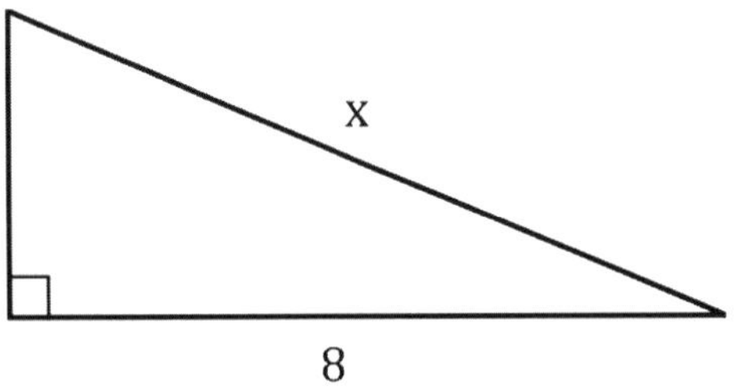

a. $A(x) = 4x$

b. $A(x) = \frac{x\sqrt{64-x^2}}{2}$

c. $A(x) = 4\sqrt{x^2 - 64}$

d. $A(x) = 64 - \sqrt{x^2}$

79. The university cafeteria offers three options:

Soup & Sandwich	Hamburger & Fries	Pizza
$7.50	$6	$5

180 students purchased meals for a total of $1,152. The number of students who bought pizza was half the number who bought the other two meals. If x, y, and z represent the number of $7.50, $6, and $5 meals sold, respectively, which of the following can be used to find how many of each sold?

a. $\begin{bmatrix} 1 & 1 & 1 \\ 7.5 & 6 & 5 \\ 2 & 2 & 1 \end{bmatrix} \begin{bmatrix} x \\ y \\ z \end{bmatrix} = \begin{bmatrix} 180 \\ 1{,}152 \\ 0 \end{bmatrix}$

b. $\begin{bmatrix} 1 & 1 & 1 \\ 7.5 & 6 & 5 \\ 1 & 1 & -2 \end{bmatrix} \begin{bmatrix} 180 \\ 1{,}152 \\ 0 \end{bmatrix} = \begin{bmatrix} x \\ y \\ z \end{bmatrix}$

c. $\begin{bmatrix} 1 & 1 & 1 \\ 7.5 & 6 & 5 \\ 1 & 1 & -2 \end{bmatrix} \begin{bmatrix} x \\ y \\ z \end{bmatrix} = \begin{bmatrix} 180 \\ 1{,}152 \\ 0 \end{bmatrix}$

d. $\begin{bmatrix} 1 & 7.5 & 1 \\ 1 & 6 & 1 \\ 1 & 5 & -2 \end{bmatrix} \begin{bmatrix} 180 \\ 1{,}152 \\ 0 \end{bmatrix} = \begin{bmatrix} x \\ y \\ z \end{bmatrix}$

80. A colony of Escherichia coli is inoculated from a Petri dish into a test tube containing 50 mL of nutrient broth. The test tube is placed in a 37 °C incubator and agitator. After one hour, the number of bacteria in the test tube is determined to be 8×10^6. Given that the doubling time of Escherichia coli is 20 minutes with agitation at 37 °C, approximately how many bacteria should the test tube contain after eight hours of growth?

a. 2.56×10^8
b. 2.05×10^9
c. 1.7×10^{14}
d. 1.7×10^{13}

Refer to the following for questions 81–83:

This data shows the number of students at a high school who are enrolled in various electives.

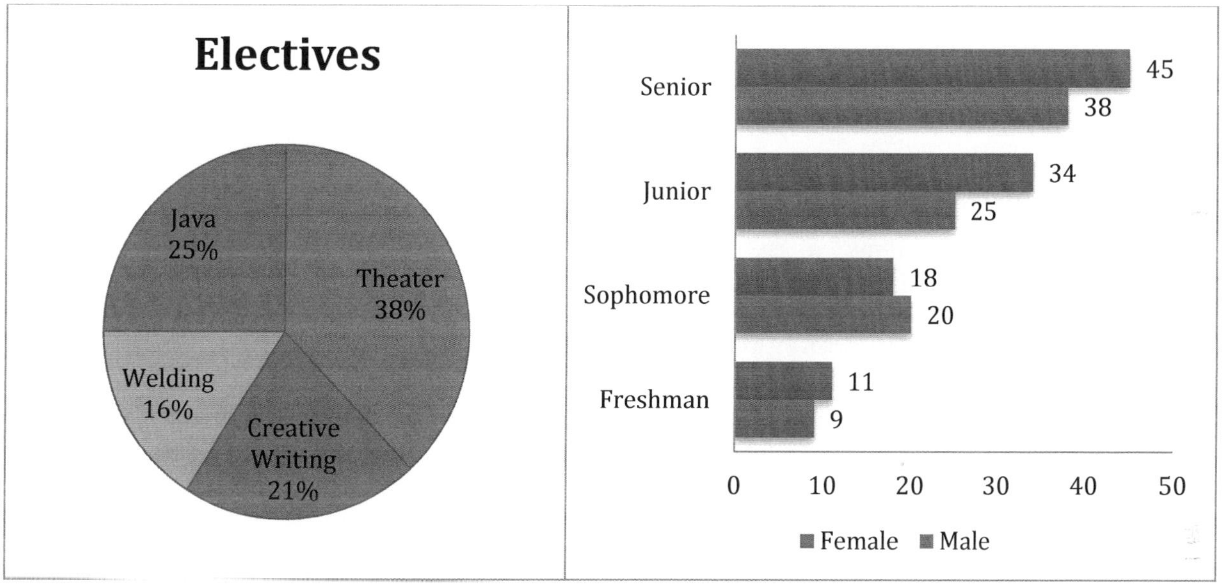

81. Which of these is the greatest quantity?

a. The number of students taking welding
b. The number of male sophomores taking an elective
c. The difference in the number of male and female students taking an elective
d. The difference in the number of juniors and seniors taking an elective

82. Compare the two quantities.

Quantity A	Quantity B
The percentage of students taking Java or Creative Writing	The percentage of students in electives who are male

a. Quantity A is greater.
b. Quantity B is greater.
c. The two quantities are the same.
d. The relationship cannot be determined from the given information.

83. A sophomore is randomly selected to be interviewed for the school paper. What is the approximate probability that the student is male?

a. 0.53
b. 0.47
c. 0.19
d. 0.10

84. For vector $v = (4, 3)$ and vector $w = (-3, 4)$, find $2(v + w)$.

a. $(2, 14)$
b. $(14, -2)$
c. $(1,7)$
d. $(7, -1)$

85. Which of these does NOT simulate randomly selecting a person from a group of 11?

a. Assigning each student a unique card value of A, 2, 3, 4, 5, 6, 7, 8, 9, 10, or J and drawing a single card from a deck of cards, redrawing if an invalid card is selected.
b. Assigning each student a unique number 2–12 and rolling a pair of dice to determine which student is selected.
c. Assigning each student a unique letter A–K, labeling 11 index cards A–K, and blindly drawing one.
d. Assigning each student a unique numeric value of 1–11 and spinning a spinner mounted on a 11-section wheel.

86. Given that x is a prime number and the greatest common factor of x and y is greater than 1, compare the two quantities.

Quantity A	Quantity B
y	the least common multiple of x and y

a. Quantity A is greater.
b. Quantity B is greater.
c. The two quantities are the same.
d. The relationship cannot be determined from the given information.

87. An adjunct professor teaches three classes. The number of women and men in each class are shown below. Two individuals are chosen at random to attend a conference with the professor. What is the approximate probability that two men from the same class will be chosen?

	Class 1	Class 2	Class 3
Women	9	21	17
Men	13	14	19

a. 7.9%
b. 9.6%
c. 11.9%
d. 13.8%

88. If $f(x) = 4x^3 - x^2 - 4x + 2$, which of the following statements is (are) true of its graph?

I. The point $\left(-\frac{1}{2}, 3\frac{1}{4}\right)$ is a relative maximum.

II. The graph of f is concave upward on the interval $(-\infty, \frac{1}{12})$.

a. I
b. II
c. I and II
d. Neither I nor II

89. A weight suspended on a spring is at its equilibrium point five inches above the top of a table. When the weight is pulled down two inches, it bounces above the equilibrium point and returns to the point from which it was released in one second. Which of these can be used to model the weight's height h above the table as a function of time t in seconds?

a. $h = -2\cos(2\pi t) + 5$
b. $h = 5\sin(\pi t) - 2$
c. $h = -2\sin(2\pi t) + 5$
d. $h = -2\cos(0.5\pi t) + 3$

90. Chan receives a bonus from his job. He pays 30% in taxes, gives 30% to charity, and uses another 25% to pay off an old debt. He has $600 left. What was the total amount of Chan's bonus?

a. $4,000
b. $3,600
c. $3,200
d. $3,000

91. Simplify the expression $\sqrt{\frac{-54x^5}{80y^6}}$.

a. $\frac{3x^2 i\sqrt{6x}}{20y^3}$
b. $\frac{3x^2 i\sqrt{6x}}{4y^3}$
c. $\frac{-3x^2 i\sqrt{30x}}{8y^3}$
d. $\frac{3x^2 i\sqrt{30x}}{20y^3}$

92. Which of these represents the equation of a sphere which is centered in the xyz-space at the point $(-1, 1, 0)$ and which has a volume of 288π cubic units?

a. $x^2 + y^2 + z^2 + 2x - 2y = 38$
b. $x^2 + y^2 + z^2 - 2x + 2y = 38$
c. $x^2 + y^2 + z^2 - 2x + 2y = 34$
d. $x^2 + y^2 + z^2 + 2x - 2y = 34$

93. Tessa rode her bike to the lake in an hour and 15 minutes. On her way home, she rode 4 mph faster and arrived in 45 minutes. What was her average speed on her round trip?

a. 3.5 mph
b. 5.5 mph
c. 7.5 mph
d. 9.5 mph

94. Roxana walks x meters west and $x + 20$ meters south to get to her friend's house. On a neighborhood map which has a scale of 1 cm : 10 m, the direct distance between Roxana's house and her friend's house is 10 cm. How far did Roxana walk to her friend's house?

a. 60 m
b. 80 m
c. 100 m
d. 140 m

95. Which of these would best illustrate the percentage of a budget allocated to various departments?

a. Pie chart
b. Line graph
c. Box-and-whisker plot
d. Venn diagram

96. Zeke drove from his house to a furniture store in Atlanta and then back home along the same route. It took Zeke three hours to drive to the store. By driving an average of 20 mph faster on his return trip, Zeke was able to save an hour of driving time. What was Zeke's average driving speed on his round trip?

a. 24 mph
b. 48 mph
c. 50 mph
d. 60 mph

97. What is the converse of the statement below?

If I see a robin, it is spring.

a. If it is not spring, I do not see a robin.
b. If it is spring, I do not see a robin.
c. If I do not see a robin, it is not spring.
d. If it is spring, I see a robin.

98. To test the accuracy and precision of two scales, a student repeatedly measured the mass of a 10g standard and recorded these results.

	Trial 1	Trial 2	Trial 3	Trial 4
Scale 1	9.99 g	9.98 g	10.02g	10.01g
Scale 2	10.206 g	10.209 g	10.210 g	10.208 g

Which of these conclusions about the scales is true?

a. Scale 1 has an average percent error of 0.15%, and Scale 2 has an average percent error of 2.08%. Scale 1 is more accurate and precise than Scale 2.

b. Scale 1 has an average percent error of 0.15%, and Scale 2 has an average percent error of 2.08%. Scale 1 is more accurate than Scale 2; however, Scale 2 is more precise.

c. Scale 1 has an average percent error of 0%, and Scale 2 has an average percent error of 2.08%. Scale 1 is more accurate and precise than Scale 2.

d. Scale 1 has an average percent error of 0%, and Scale 2 has an average percent error of 2.08%. Scale 1 is more accurate than Scale 2; however, Scale 2 is more precise.

99. A private tutor works with 11 five-year-olds, 8 six-year-olds, 9 seven-year-olds, 14 eight-year-olds, 11 nine-year-olds, and 13 ten-year-olds. If a student is randomly selected for a free session, what is the probability that he or she is no more than eight years old?

a. $\frac{7}{11}$

b. $\frac{14}{33}$

c. $\frac{4}{11}$

d. $\frac{3}{7}$

100. Which of the following could be the graph of $y = a(x+b)^2(x+c)^2$ if $a > 0$?

a.

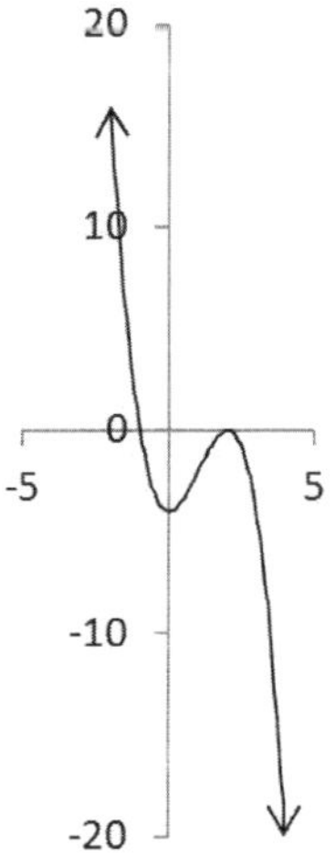

b.

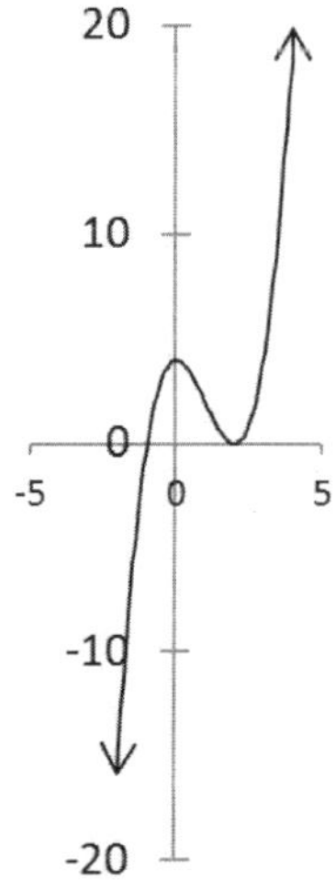

c.

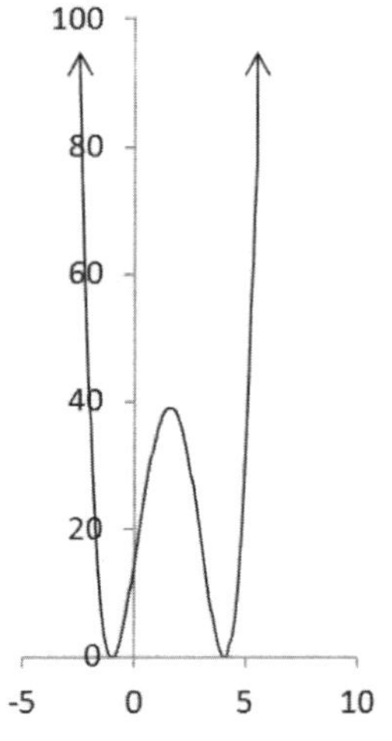

d.

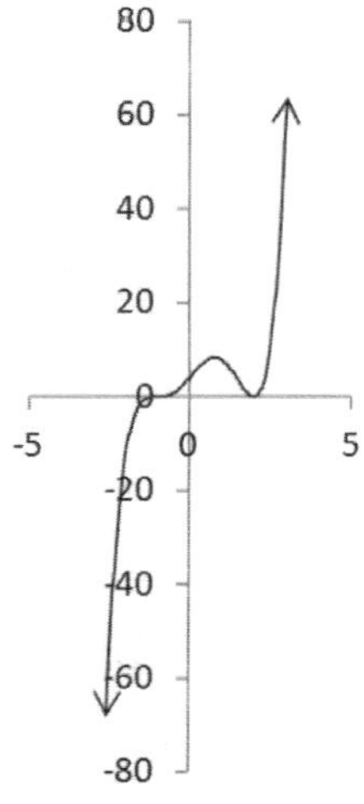

Answer Key and Explanations

1. B: The recursive definition of the sequence gives the first term of the series, $a_1 = 2$. The definition also defines each term in the series as the sum of the previous term and -3. Therefore, the second term in the series is $2 + (-3) = -1$, the third term in the series is $-1 + (-3) = -4$, and so on.

n	a_n
1	2
2	-1
3	-4

The relationship between n and a_n is linear, so the equation of the sequence can be found using the explicit formula for an arithmetic sequence, $a_n = a_1 + d(n-1)$, where a_n is the value of any term n, a_1 is the value of the first term, and d is the common difference between each term. The value of a_n decreases by 3 each time the value of n increases by 1, so the common difference (d) is -3.

Substitute all the known values into the equation for the explicit formula for an arithmetic sequence, and simplify the formula.

$$a_n = 2 + (-3)(n-1)$$
$$a_n = 2 - 3n + 3$$
$$a_n = -3n + 5$$

2. C: Will used the following calculations.

$$1 \text{ bag of flour } \times \frac{5 \text{ lb}}{1 \text{ bag}} \times \frac{16 \text{ oz}}{1 \text{ lb}} \times \frac{25 \text{ g protein}}{4 \text{ oz}} \times \frac{1 \text{ serving}}{50 \text{ g protein}} \approx \frac{5 \times 16 \times 25}{4 \times 50} = 10$$

So, Will should divide the flour into 10 equal portions to equal the weight of chicken that would contain 50 grams of protein.

3. C: The sequence consists of the squares of consecutive integers, beginning with 1: $1^2 = 1$, $2^2 = 4$, $3^2 = 9$, and $4^2 = 16$. In the fifth position, $5^2 = 25$, so 25 is the next number.

4. B: The average rate of cooling is: (86° - 38°) / 9 hrs; 48° / 9 = 5.33° F per hour.

5. A: The area of a circle is πr^2, so the area of a semicircle is $\frac{\pi r^2}{2}$. Illustrated below is a method that can be used to find the area of the shaded region.

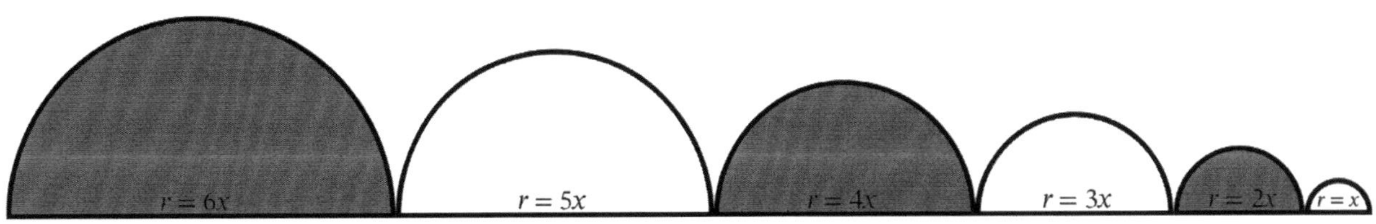

$$\begin{aligned} A &= \frac{\pi(6x)^2}{2} - \frac{\pi(5x)^2}{2} + \frac{\pi(4x)^2}{2} - \frac{\pi(3x)^2}{2} + \frac{\pi(2x)^2}{2} - \frac{\pi(x)^2}{2} \\ &= \frac{36\pi x^2}{2} - \frac{25\pi x^2}{2} + \frac{16\pi x^2}{2} - \frac{9\pi x^2}{2} + \frac{4\pi x^2}{2} - \frac{\pi x^2}{2} \\ &= \frac{(36 - 25 + 16 - 9 + 4 - 1)\pi x^2}{2} \\ &= \frac{21\pi x^2}{2} \end{aligned}$$

6. C: The volume of the paperweight is $\frac{x^2z}{3}$. If each side of the square base were increased by $\frac{1}{2}$ in, the volume would be $\frac{\left(x+\frac{1}{2}\right)^2 z}{3} = \frac{\left(x^2+x+\frac{1}{4}\right)(z)}{3} = \frac{x^2z+xz+\frac{1}{4}z}{3} = \frac{x^2z}{3} + \frac{4xz+z}{12}$. So, the original volume of the paperweight, $\frac{x^2z}{3}$, increased by $\frac{4xz+z}{12}$, or $\frac{z(4x+1)}{12}$.

7. C: In this sequence, the interval between successive numbers increases by an amount that is one less than the previous interval.

$$23 - 18 = 5$$
$$27 - 23 = 4$$
$$33 - 32 = 1$$

Therefore, the next number should be $27 + 3$, or 30.

8. D: Integers consist of all positive and negative whole numbers and the number zero. The product of three integers must be an integer, so you can eliminate any answer choice that is not an integer. The product of two even integers is even. The product of an even and odd integer is even. The only even choice is 24, and we can see that $2 \times 4 \times 3 = 24$.

9. A: The average number of male students in the 11th and 12th grades is $\frac{131 + 119}{2} = 125$. The number of Hispanic students at the school is 10% of 1,219, which is $0.10 \times 1{,}219 = 121.9 \approx 122$ students. The difference in the number of male and female students at the school is $630 - 589 = 41$, and the difference in the number of 9th and 12th grade students at the school is $354 - 255 = 99$.

10. D: 131 of 283 eleventh graders are male. It is given that an eleventh grader is chosen to attend the conference, so the probability that a male is chosen is $\frac{\text{number of 11th grade males}}{\text{number of 11th graders}} = \frac{131}{283} \approx 0.46$.

11. C: 52% of the student population is Caucasian. There are 630 female students at the school out of 1,219 students, so the percentage of female students is $\frac{630}{1{,}219} \times 100\% \approx 52\%$. The percentages, rounded to the nearest whole number, are the same.

12. B: If each of the 4 groups in the class of 20 will contain 3 boys and 2 girls, there must be 12 boys and 8 girls in the class. The number of ways the teacher can select 3 boys from a group of 12 boys is:

$$_{12}C_3 = \frac{12!}{3!\,(12-3)!} = \frac{12!}{3!\,9!} = \frac{12 \times 11 \times 10 \times 9!}{3!\,9!} = \frac{12 \times 11 \times 10}{3 \times 2 \times 1} = 220$$

The number of ways she can select 2 girls from a group of 8 girls is:

$$_{8}C_2 = \frac{8!}{2!\,(8-2)!} = \frac{8!}{2!\,6!} = \frac{8 \times 7 \times 6!}{2!\,6!} = \frac{8 \times 7}{2 \times 1} = 28$$

Since each combination of boys can be paired with each combination of girls, the number of group combinations is $220 \times 28 = 6{,}160$.

13. B: The endpoint and midpoint of the line segment must lie on the same line. Since we are given the slope of a line and a point on the line, use the point-slope form equation to find the equation of the line.

$$y - y_1 = m(x - x_1)$$
$$y - 3 = -\frac{1}{2}\big(x - (-2)\big)$$
$$y - 3 = -\frac{1}{2}(x + 2)$$
$$y - 3 = -\frac{1}{2}x - 1$$
$$y = -\frac{1}{2}x + 2$$

From here, substitute the given points into the equation to determine which one is a possible endpoint of the line segment.

a. $(-4, 3)$	b. $(2, 1)$	c. $(-6, 6)$	d. $\left(0, \frac{3}{2}\right)$
$y = -\frac{1}{2}x + 2$	$y = -\frac{1}{2}x + 2$	$y = -\frac{1}{2}x + 2$	$y = -\frac{1}{2}x + 2$
$3 = -\frac{1}{2}(-4) + 2$	$1 = -\frac{1}{2}(2) + 2$	$6 = -\frac{1}{2}(-6) + 2$	$\frac{3}{2} = -\frac{1}{2}(0) + 2$
$3 = 2 + 2$	$1 = -1 + 2$	$6 = 3 + 2$	$\frac{3}{2} = 0 + 2$
$3 = 4$	$1 = 1$	$6 = 5$	$\frac{3}{2} = 2$

The only point that results in a true statement is (2,1), so this is the only point that could be a possible endpoint of the line segment.

14. B: This is the graph of the parent function $y = \sqrt{x}$. Another way to determine the function is by testing points. Notice that the graph passes through the points (0,0), and (4,2). A quick check of each x-value shows that the only function that produces y-values of 0 and 2 is $y = \sqrt{x}$.

x	x^2	$\sqrt{x}$	2^x	$\log_2 x$
0	0	0	1	undefined
4	16	2	16	2

15. A: The Babylonians used a base-60 numeral system, which is still used in the division of an hour into 60 minutes, a minute into 60 seconds, and a circle into 360 degrees. The word "algebra" and its development as a discipline separate from geometry are attributed to the Arabic/Islamic civilization. The Greek philosopher Thales is credited with using deductive reasoning to prove geometric concepts. Boolean logic was introduced by British mathematician George Boole.

16. A: First, subtract the matrices in parentheses, then multiply the resulting matrices together.

$$\begin{aligned}[2 \quad 0 \quad -5]\begin{bmatrix}(4-3)\\(2-5)\\(-1-(-5))\end{bmatrix} &= [2 \quad 0 \quad -5]\begin{bmatrix}1\\-3\\4\end{bmatrix}\\ &= [(2)(1)+(0)(-3)+(-5)(4)]\\ &= [-18]\end{aligned}$$

Note that the product of a 1×3 matrix and a 3×1 matrix, in that order, is a 1×1 matrix. If the order was reversed, the product would be a 3×3 matrix.

17. D: To draw a box-and-whisker plot from the data, find the median, quartiles, and upper and lower limits.

```
3|6 799
4|2|3889             Key
5|0|11158|   2|123 = 21, 22, 23
6|0 0123
```

The median is $\frac{50+51}{2} = 50.5$, the lower quartile is $\frac{42+43}{2} = 42.5$, and the upper quartile is $\frac{58+60}{2} = 59$. The box of the box-and-whisker plot goes through the quartiles, and a line through the box represents the median of the data. The whiskers extend from the box to the lower and upper limits, unless there are any outliers in the set. In this case, there are no outliers, so the box-and-whisker plot in choice A correctly represents the data set.

To draw a pie chart, find the percentage of data contained in each of the ranges shown. There are four out of twenty numbers between 30 and 39, inclusive, so the percentage shown in the pie chart for that range of data is $\frac{4}{20} \times 100\% = 20\%$; there are five values between 40 to 49, inclusive, so the percentage of data for that sector is $\frac{5}{20} \times 100\% = 25\%$; $\frac{6}{20} \times 100\% = 30\%$ of the data is within the range of 50–59, and $\frac{5}{20} \times 100\% = 25\%$ is within the range of 60–69. The pie chart shows the correct percentage of data in each category.

To draw a cumulative frequency histogram, find the cumulative frequency of the data.

Range	Frequency	Cumulative frequency
30–39	4	4
40–49	5	9
50–59	6	15
60–69	5	20

The histogram shows the correct cumulative frequencies. Therefore, all of the graphs represent the data set.

18. D: Solve this equation by completing the square or by using the quadratic formula, $x = \frac{-b\pm\sqrt{b^2-4ac}}{2a}$. Here we will use the quadratic formula. First, set the equation equal to 0 by adding 2 to both sides. The equation $5x^2 - 4x + 2 = 0$ has $a = 5, b = -4, c = 2$. Substitute these values into the quadratic formula and simplify.

$$x = \frac{-b \pm \sqrt{b^2 - 4ac}}{2a}$$
$$x = \frac{4 \pm \sqrt{(-4)^2 - 4(5)(2)}}{2(5)}$$
$$x = \frac{4 \pm \sqrt{16 - 40}}{10}$$
$$x = \frac{4 \pm \sqrt{-24}}{10}$$
$$x = \frac{4 \pm 2i\sqrt{6}}{10}$$
$$x = \frac{2 \pm i\sqrt{6}}{5}$$

19. B: First, state the exclusions of the domain. This can be done by factoring the denominator, then finding where the denominator is equal to 0.

$$x^3 + 2x^2 - x - 2 \neq 0$$
$$x^2(x + 2) - (x + 2) \neq 0$$
$$(x + 2)(x^2 - 1) \neq 0$$
$$(x + 2)(x - 1)(x + 1) \neq 0$$
$$x + 2 \neq 0 \quad x - 1 \neq 0 \quad x + 1 \neq 0$$
$$x \neq -2 \quad x \neq 1 \quad x \neq -1$$

To determine whether there are vertical asymptotes or holes at these values of x, simplify the expression:

$$\frac{x^2 - x - 6}{x^3 + 2x^2 - x - 2} = \frac{(x - 3)(x + 2)}{(x + 2)(x - 1)(x + 1)} = \frac{x - 3}{(x - 1)(x + 1)}$$

There are vertical asymptotes at $x = 1$ and at $x = -1$ and a hole at $x = -2$. Statement I is false.

To find the x-intercept of $f(x)$, solve $f(x) = 0$. $f(x) = 0$ when the numerator is equal to zero.

$$\begin{aligned} x^2 - x - 6 &= 0 \\ (x + 2)(x - 3) &= 0 \\ x &= -2,3 \end{aligned}$$

The numerator equals zero when $x = -2$ and $x = 3$; however, -2 is excluded from the domain of $f(x)$, so the x-intercept is 3. To find the y-intercept of $f(x)$, find $f(0)$.

$$\frac{(0)^2 - (0) - 6}{(0)^3 + 2(0)^2 - (0) - 2} = \frac{-6}{-2} = 3$$

The y-intercept is 3. Statement II is true.

20. C: When the shirt is marked down by 25%, the cost of the shirt is 75% of its original price. Therefore, the reduced price of the shirt can be written as $\frac{75}{100}x$, or $\frac{3}{4}x$, where x is the original price. When discounted an extra 30%, the shirt costs 70% of the reduced price, or $\frac{70}{100}\left(\frac{3}{4}x\right)$, or $\frac{7}{10}\left(\frac{3}{4}x\right)$, which simplifies to $\frac{21}{40}x$. So, the final price is $\frac{21}{40}$ of the original price.

21. B: The only cross-section connecting the points that is not possible is a square; a rectangle is possible, but a square is not.

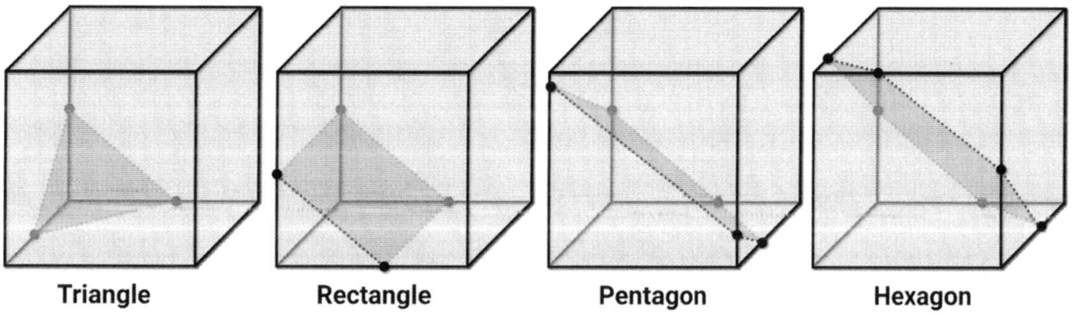

22. B: Use a Venn diagram to help organize the given information. Start by filling in the space where the three circles intersect: six customers ordered all three items. Now, use that information to fill in the spaces where two circles intersect. For example, fourteen customers ordered both pumpkin bread and cranberry muffins, and six of those were the customers who purchased all three items, so eight customers bought pumpkin bread and cranberry muffins but not eggnog pie. Once the diagram is completed, add the number of orders from each portion of the diagram. The total number of customers was $34 + 19 + 51 + 9 + 8 + 4 + 6 = 131$.

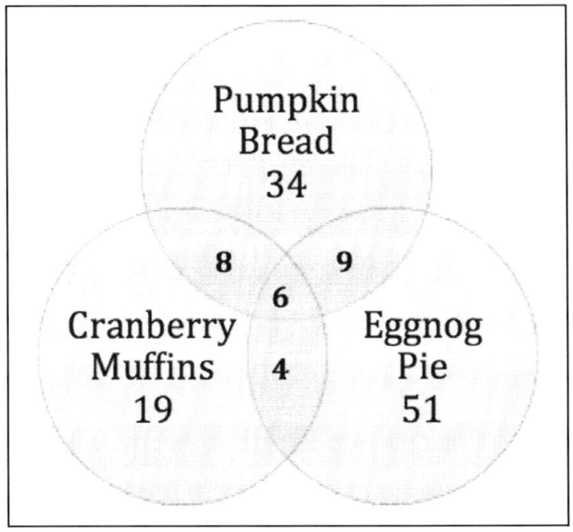

23. A: Below is the graph of $g(x)$.

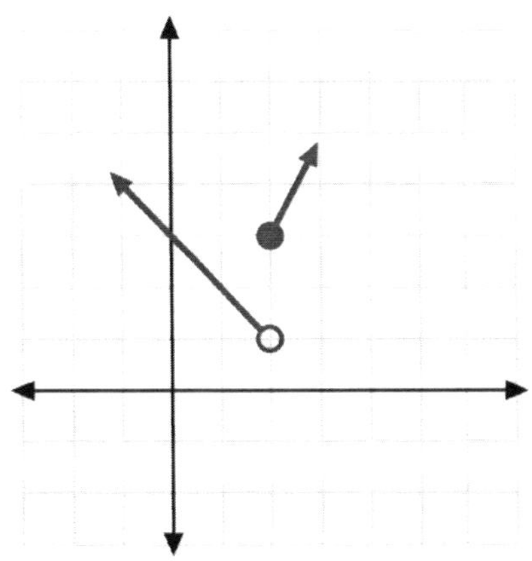

Statement II is true. The graph is discontinuous at $x = 2$. Statement I is false, because $g(3) = 2(3) - 1 = 5$. Statement III is also false because, from the graph, the range of the function is $y > 1$.

24. D: Since it is given that $\overline{FD} \cong \overline{BC}$ and $\overline{AB} \cong \overline{DE}$, step 2 needs to establish either that $\overline{AC} \cong \overline{EF}$ or that $\Delta ABC \cong \Delta FDE$ in order for step 5 to show that $\Delta ABC \cong \Delta EDF$. The statement $\overline{AC} \cong \overline{EF}$ cannot be shown directly from the given information. On the other hand, $\Delta ABC \cong \Delta FDE$ can be determined: when two parallel lines $\overline{BC} \parallel \overline{FG}$ are cut by a transversal ($\overline{AE}$), alternate exterior angles ($\Delta ABC, \Delta FDE$) are congruent. Therefore, $\Delta ABC \cong \Delta EDF$ by the side-angle-side (SAS) theorem.

25. A: Step 5 established that $\Delta ABC \cong \Delta EDF$. Because corresponding parts of congruent triangles are congruent (CPCTC), $\angle BAC \cong \angle DEF$. This is useful to establish when trying to prove $\overline{FE} \parallel \overline{AG}$: when two lines ($\overline{FE}$ and $\overline{AG}$) are cut by a transversal ($\overline{AE}$) and alternate interior angles ($\angle BAC$, $\angle DEF$) are congruent, then the lines are parallel. The completed proof is shown below:

Statement	Reason
1. $\overline{BC} \parallel \overline{FG}$	Given

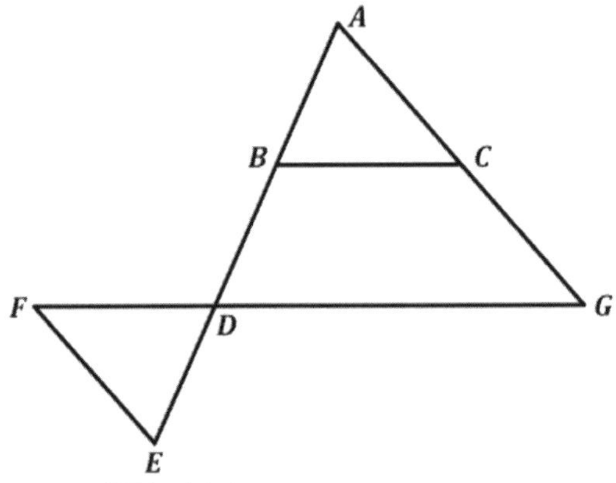

Given: $\overline{BC} \parallel \overline{FG}$; $\overline{FD} \cong \overline{BC}$; $\overline{AB} \cong \overline{DE}$
Prove: $\overline{FE} \parallel \overline{AG}$

Statement	Reason
2. $\angle ABC \cong \angle FDE$	Alternate exterior angles of parallel lines are congruent
3. $\overline{FD} \cong \overline{BC}$	Given
4. $\overline{AB} \cong \overline{DE}$	Given
5. $\Delta ABC \cong \Delta EDF$	SAS
6. $\angle BAC \cong \angle DEF$	CPCTC
7. $\overline{FE} \parallel \overline{AG}$	Alternate interior angles congruent then lines are parallel

■

26. A: A triangular pyramid has four triangular faces. The arrangement of these faces in a two-dimensional figure is a net of a triangular pyramid if the figure can be folded to form a triangular pyramid. Choice B represents a rectangular pyramid, choice C is a triangular prism, and choice D is a square pyramid.

27. A: A coterminal angle for $-\frac{\pi}{4}$ is $2\pi - \frac{\pi}{4} = \frac{7\pi}{4}$.

$$\tan\left(-\frac{\pi}{4}\right) = \tan\left(\frac{7\pi}{4}\right) = \frac{\sin\left(\frac{7\pi}{4}\right)}{\cos\left(\frac{7\pi}{4}\right)}$$

From the unit circle, the values of $\sin\left(\frac{7\pi}{4}\right)$ and $\cos\left(\frac{7\pi}{4}\right)$ are $-\frac{\sqrt{2}}{2}$ and $\frac{\sqrt{2}}{2}$, respectively. Therefore:

$$\tan\left(-\frac{\pi}{4}\right) = \frac{-\frac{\sqrt{2}}{2}}{\frac{\sqrt{2}}{2}} = -1$$

28. A: This problem is most easily solved using the law of sines, which states that the ratio of the sine of each angle in a triangle to the length of the opposite side is equal: $\frac{\sin A}{a} = \frac{\sin B}{b} = \frac{\sin C}{c}$. In this case, angle θ is opposite side a, and the angle with a measure of 30° is opposite side b, so we can write $\frac{\sin\theta}{a} = \frac{\sin 30°}{b}$. Since $\sin 30° = \frac{1}{2}$, this becomes $\frac{\sin\theta}{a} = \frac{\frac{1}{2}}{b}$, or $\sin\theta = \frac{a}{2b}$.

29. D: Since there are 100 homes' market times represented in each set, the median time a home spends on the market is between the 50th and 51st data point in each set. The 50th and 51st data points for Zip Code 1 are six months and seven months, respectively, so the median time a house in Zip Code 1 spends on the market is between six and seven months (6.5 months), which by the realtor's definition of market time is a seven-month market time. The 50th and 51st data points for Zip Code 2 are both thirteen months, so the median time a house in Zip Code 2 spends on the market is thirteen months. Therefore, there is a six-month difference between the median of Zip Code 1 and the median of Zip Code 2.

To find the mean market time for 100 houses, find the sum of the market times and divide by 100. If the frequency of a one-month market time is 9, the number 1 is added nine times (1×9), if the frequency of a two-month market time is 10, the number 2 is added ten times (2×10), and so on. So, to find the average market time, divide by 100 the sum of the products of each market time and its corresponding frequency. For Zip Code 1, the mean market time is 7.38 months, which, by the realtor's definition of market time, is an eight-month market time. For Zip Code 2, the mean market time is 12.74, which, by the realtor's definition of market time, is a thirteen-month market time. Therefore, the difference between the mean of Zip Code 1 and the mean of Zip Code 2 is 5 months.

The mode market time is the market time for which the frequency is the highest. For Zip Code 1, the mode market time is three months, and for Zip Code 2, the mode market time is eleven months. Therefore, the median time a house spends on the market in Zip Code 1 is less than the mean time a house spends on the market in Zip Code 1.

30. C: "The square of twice the sum of x and three is equal to the product of twenty-four and x" is represented by the equation $[2(x+3)]^2 = 24x$. Solve for x.

$$\begin{aligned}
[2x+6]^2 &= 24x \\
(2x+6)(2x+6) &= 24x \\
4x^2 + 24x + 36 &= 24x \\
4x^2 + 36 &= 0 \\
4x^2 &= -36 \\
x^2 &= -9 \\
x &= \pm\sqrt{-9} \\
x &= \pm 3i
\end{aligned}$$

So, $-3i$ is a possible value of x.

31. A: The ratio of expenses to income is 7: 1. Since we know he saves \$725 per month, we need to multiply this value by 7 to find out how much he spends on expenses each month. $\$725 \times 7 = \$5{,}075$. So, Jon spends \$5,075 on expenses each month.

32. D: Since the graph shows a maximum height of 24 inches above the ground, and since the maximum distance from the road the pebble reaches occurs when it is at the top of the tire, the diameter of the tire is 24 inches. Therefore, its radius is 12 inches. From the graph, it can be observed that the tire makes 6 rotations in 2 seconds. Thus, the tire rotates 3 times in 1 second, or $3 \times 60 = 180$ times per minute.

33. C: The situation may be modeled by the system of inequalities $\begin{cases} 6x + 3y \le 75 \\ x + y \le 30 \end{cases}$ where x is the number of coffee mugs and y is the number of note pads. Some algebraic manipulation gives us the inequalities in slope-intercept form: $\begin{cases} y \le -2x + 25 \\ y \le -x + 30 \end{cases}$. All four choices graph these lines correctly, but only choice C correctly shades just the region that is below both lines, indicating that both conditions are met.

34. D: The graph of $f(x)$ is stretched vertically by a factor of 2 and is inverted with respect to $y = \sin x$, so $a_1 = -2$. The graph of $g(x)$ is stretched vertically by a factor of 4 with respect to the graph of $y = \cos x$, so $a_2 = 4$. Therefore, the statement $a_1 < 0 < a_2$ is true.

35. C: There are $(n-1)!$ ways to arrange n units around a circle for $n > 1$. Because the teacher and Andrew must sit together, we can treat them as a unit, as though there were five people. So we calculate:

$$(n-1)! = (5-1)! = 4! = 24$$

There are 24 possible arrangements with Andrew and the teacher counted as a single unit. But for each of those, Andrew could be sitting either to her left or to her right, so we double the amount: $24 \times 2 = 48$.

36. C: There are nine ways to assign the first digit since it can be any of the numbers 1-9. There are nine ways to assign the second digit since it can be any digit 0-9 except for the digit assigned in place 1. There are eight ways to assign the third number since there are ten digits, two of which have already been assigned. There are seven ways to assign the fourth number, six ways to assign the fifth, five ways to assign the sixth, and four ways to assign the seventh. Multiply to find the total number of combinations.

$$9 \times 9 \times 8 \times 7 \times 6 \times 5 \times 4 = 544{,}320.$$

Another way to approach the problem is to notice that the arrangement of nine digits in the last six places is a sequence without repetition, or a permutation. (Note: this may be called a partial permutation since all of the elements of the set need not be used.) The number of possible sequences of a fixed length r of elements taken from a given set of size n is permutation ${}_nP_r = \frac{n!}{(n-r)!}$. So, the number of ways to arrange the last six digits is ${}_9P_6 = \frac{9!}{(9-6)!} = \frac{9!}{3!} = 60{,}480$. Multiply this number by nine since there are nine possibilities for the first digit of the phone number: ${}_9P_6 \times 9 = 544{,}320$.

37. C: Sketch a diagram and label the known segments. Use the property that two segments are congruent when they originate from the same point outside of a circle and are tangent to the circle. The point of tangency of $\overline{CD}$ divides the segment into two pieces measuring 4 and 4. The point of tangency of $\overline{BC}$ divides the segment into two pieces measuring 4 and 8; the point of tangency of $\overline{AB}$ divides the segment into two pieces measuring 2 and 8. Finally, the point of tangency of $\overline{AD}$ divides the segment into two pieces measuring 2 and 4. Therefore, $\overline{AD} = 2 + 4 = 6$.

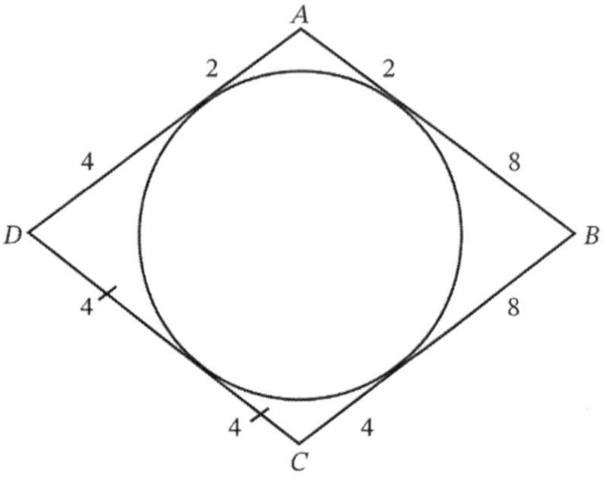

38. D: The probability of an event is the number of possible occurrences of that event divided by the number of all possible outcomes. A student who is at least nine years old can be nine or ten years old, so the probability of randomly selecting a student at least nine years old is:

$$\frac{\text{number of nine- and ten-year old students}}{\text{total number of students}} = \frac{10+14}{10+9+13+10+14} = \frac{24}{56} = \frac{3}{7}$$

39. A: One way to determine which is the graph of $f(x) = -2|-x+4| - 1$ is to find $f(x)$ for a few values of x. For example, when $x = 0$, $f(0) = -2|0+4| - 1 = -9$. Graphs A and B pass through $(0,-9)$, but graphs C and D do not. When $x = 4$, $f(4) = -2|-4+4| - 1 = -1$. Graph A passes through $(4,-1)$, but graph B does not. Graph A is the correct graph.

40. A: First, write three equations from the information given in the problem. Since the total number of tickets sold was 810, $x + y + z = 810$. The ticket sales generated \$14,500, so $15x + 25y + 20z = 14{,}500$. The number of children under ten was the same as twice the number of adults and seniors, so $x = 2(y+z)$, which can be rewritten as $x - 2y - 2z = 0$.

The coefficients of each equation are arranged in the rows of a 3×3 matrix, which, when multiplied by the 3×1 matrix arranging the variables x, y, and z, will give the 3×1 matrix which arranges the constants of the equations.

41. A: First, multiply the numerator and denominator by the denominator's conjugate, $4 + 2i$. Then, simplify the result and write the answer in the form $a + bi$. Remember, $i^2 = -1$.

$$\begin{aligned}\frac{2+3i}{4-2i} \times \frac{4+2i}{4+2i} &= \frac{8+4i+12i+6i^2}{16-4i^2}\\ &= \frac{8+16i-6}{16+4}\\ &= \frac{2+16i}{20}\\ &= \frac{2}{20}+\frac{16i}{20}\\ &= \frac{1}{10}+\frac{4}{5}i\end{aligned}$$

42. A: 28% of the volunteers work in construction. There are $90 + 72 = 162$ female volunteers over age 40, and the total number of volunteers is $90 + 97 + 72 + 71 + 82 + 96 + 53 + 65 = 626$. This means that the percentage of volunteers who are females over age 40 is $\frac{162}{626} \times 100\% \approx 26\%$. Therefore, quantity A is greater than quantity B.

43. B: To find the best estimate for this, start by rounding each number to the nearest whole number. To round 23.97124, look at the tenths place (9). Since this number is greater than or equal to 5, 23.97124 rounds up to 24. To round 8.023 to the nearest whole number, look again at the tenths place (0). Since it is less than 5, 8.023 rounds down to 8. Now these rounded numbers can be divided to find the best estimate. $24 \div 8 = 3$, so 3 is the best estimate for $23.79124 \div 8.023$.

44. D: The vertex form of a quadratic equation is $y = a(x-h)^2 + k$, where $x = h$ is the parabola's axis of symmetry and (h,k) is the parabola's vertex. The vertex of the graph is $(-1,3)$, so the equation can be written as $y = a(x+1)^2 + 3$. The parabola passes through point (1,1). Substitute $x = 1$ and $y = 1$ into the equation. So, $1 = a(1+1)^2 + 3$. Solve for a.

$$\begin{aligned} 1 &= a(1+1)^2 + 3 \\ 1 &= a(2)^2 + 3 \\ 1 &= 4a + 3 \\ -2 &= 4a \\ -\frac{1}{2} &= a \end{aligned}$$

The vertex form of the parabola is $y = -\frac{1}{2}(x+1)^2 + 3$. Write the equation in standard form, $y = ax^2 + bx + c$.

$$\begin{aligned} y &= -\frac{1}{2}(x+1)^2 + 3 \\ y &= -\frac{1}{2}(x^2 + 2x + 1) + 3 \\ y &= -\frac{1}{2}x^2 - x - \frac{1}{2} + 3 \\ y &= -\frac{1}{2}x^2 - x + \frac{5}{2} \end{aligned}$$

45. B: The range is the spread of the data. It can be calculated for each test by subtracting the lowest test score from the highest, or it can be determined visually from the graph. The difference between the highest and lowest test scores on the Midterm is $91 - 22 = 69$ points. The range for each of the other classes is much smaller.

46. D: For the Final, the third quartile is 87. Since 75% of the data in a set is below the third quartile, there is a 75% chance, or a probability of 0.75, that a test score chosen at random from the Final scored below 87.

47. C: If $f(x) = 10^x$ and $f(x) = 5$, then $5 = 10^x$. Write this in the form of a logarithm, $\log_{10}5 = x$. Now, use a calculator to find that $x \approx 0.7$.

48. C: Leonhard Euler made many important contributions to the field of mathematics. One such contribution, Euler's formula $e^{i\varphi} = \cos\varphi + i\sin\varphi = 0$, can be written as $e^{i\pi} + 1 = 0$ when $\varphi = \pi$.

49. B: Rounding to the nearest tenth means adjusting the first digit after the decimal to the value closest to the actual calculated answer, and then truncating the answer at that digit. Thus, $4.171 + 4.55 = 8.721$, which rounds to 8.7.

50. C: The x-intercept is the point at which $f(x) = 0$. When $0 = \log_b x$, $b^0 = x$; since $b^0 = 1$, the x-intercept of $f(x) = \log_b x$ is always 1. If $f(x) = \log_b x$ and $x = b$, then $f(x) = \log_b b$, which is, by definition, 1 ($b^1 = b$). If $g(x) = b^x$, then $f(x)$ and $g(x)$ are inverse functions and are therefore symmetric with respect to $y = x$. The statement choice C is not necessarily true since $x < 1$ includes numbers between 0 and 1. The function is undefined when $x < 0$, but not necessarily for $0 < x < 1$.

51. C: The final sales price of the couch is:

$$1.08(0.8 \times \$3{,}200) = \$2{,}764.80 \quad \text{at Store I}$$
$$\$3{,}050 + \$50 = \$3{,}100.00 \quad \text{at Store II}$$
$$1.08(\$3{,}375 - \$750) = \$2{,}835.00 \quad \text{at Store III}$$

So Store I is the least expensive, Store III is next, and Store II is the most expensive.

52. C: A pie chart shows the relationship of parts to a whole. A line graph is often used to show change over time. A box-and-whisker plot displays how numeric data are distributed throughout the range. Box-and-whisker plot is the correct answer. A Venn diagram shows the relationships among sets.

53. C: If l and w represent the length and width of the enclosed area, its perimeter is equal to $2l + 2w$; since the fence is positioned x feet from the lot's edges on each side, the perimeter of the lot is $2(l + 2x) + 2(w + 2x)$. Since the amount of money saved by fencing the smaller area is \$432, and since the fencing material costs \$12 per linear foot, 36 fewer feet of material are used to fence around the playground than would have been used to fence around the lot. This can be expressed as the following equation.

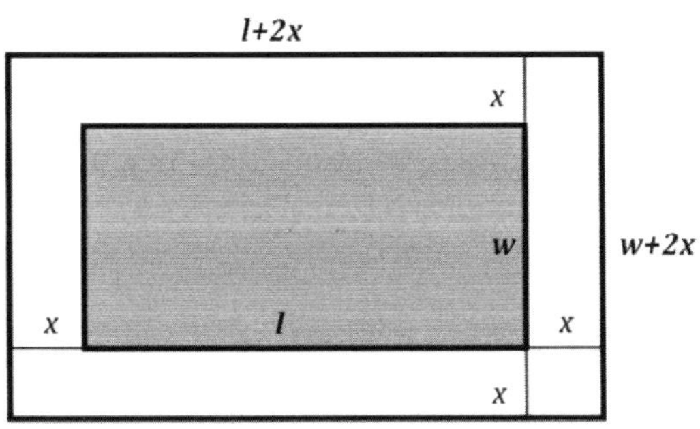

$$2(l + 2x) + 2(w + 2x) - (2l + 2w) = 36$$
$$2l + 4x + 2w + 4x - 2l - 2w = 36$$
$$8x = 36$$
$$x = 4.5 \text{ ft}$$

The difference in the area of the lot and the enclosed space is 141 yd^2, which is the same as 1,269 ft^2. So, $(l + 2x)(w + 2x) - lw = 1{,}269$. Substituting 4.5 for x, we get the following equation that we can solve for $l + w$.

$$(l + 9)(w + 9) - lw = 1{,}269$$
$$lw + 9l + 9w + 81 - lw = 1{,}269$$
$$9l + 9w = 1{,}188$$
$$9(l + w) = 1{,}188$$
$$l + w = 132 \text{ ft}$$

Therefore, the perimeter of the enclosed space, $2(l + w)$, is $2(132) = 264$ ft. The cost of 264 ft of fencing is $264 \times \$12 = \$3{,}168$.

54. B: The notation $\mathbb{P} \subseteq \mathbb{N} \subseteq \mathbb{Z} \subseteq \mathbb{Q} \subseteq \mathbb{R} \subseteq \mathbb{C}$ means that the set of prime numbers is a subset of the set natural numbers, which is a subset of the set of integers, which is a subset of the set of rational numbers, which is a subset of the set real numbers, which is a subset of the set of complex numbers. The set of prime numbers ($\mathbb{P}$) are natural numbers that have no factors other than 1 and itself. The set of natural numbers ($\mathbb{N}$) are all positive integers. The set of integers ($\mathbb{Z}$) are all positive and negative whole numbers, numbers without a fraction or decimal part to them. The set of rational numbers ($\mathbb{Q}$) are all numbers that can be expressed as the quotient of two integers. The set of real numbers ($\mathbb{R}$) are all numbers that can be expressed on a number line. The set of all complex numbers ($\mathbb{C}$) are all real and imaginary numbers.

55. D: There are 12 inches in a foot, so on a $1:24$ scale drawing, each inch represents 24 in, or 2 ft. Therefore, an area of 1 in^2 on the drawing represents $2 \text{ ft} \times 2 \text{ ft} = 4 \text{ ft}^2$ in actuality. Since the area of the room in the scale drawing is 25 in^2, the room's actual area is $25 \times 4 = 100 \text{ ft}^2$.

Another way to determine the area of the room is to write and solve an equation such as this one: $\frac{l}{24} \times \frac{w}{24} = 25 \text{ in}^2$, where l and w are the dimensions of the actual room.

$$\frac{lw}{576} = 25 \text{ in}^2$$

$$\text{Area} = 14{,}400 \text{ in}^2$$

Since this is not one of the answer choices, convert square inches to square feet.

$$14{,}400 \text{ in}^2 \times \frac{1 \text{ ft}}{12 \text{ in}} \times \frac{1 \text{ ft}}{12 \text{ in}} = 100 \text{ ft}^2$$

56. A: Calculate the final sales price of the rug at each store. Please note that the explanation below uses 1.08 instead of 0.08 for tax in order to avoid having to perform addition later. If 0.08 were to be used, only the amount of tax charged would result instead of the total cost of the product with tax included. Thus, 1.08 is used to make sure 108% of the product price is calculated.

$$1.08(0.7 \times \$296) = \$223.78 \text{ at Store A}$$
$$1.08(\$220 - \$10) = \$226.80 \text{ at Store B}$$
$$\$198 + \$35 = \$233 \text{ at Store C}$$

57. B: To solve the equation, write $x^4 + 64 = 20x^2$ in the quadratic form, then factor.

$$(x^2)^2 - 20(x^2) + 64 = 0$$

This trinomial can be factored as $(x^2 - 4)(x^2 - 16) = 0$. Each set of parentheses contains a difference of squares, which can be factored further: $(x + 2)(x - 2)(x + 4)(x - 4) = 0$. Use the zero product property to find the solutions to the equation.

$x + 2 = 0$	$x - 2 = 0$	$x + 4 = 0$	$x - 4 = 0$
$x = -2$	$x = 2$	$x = -4$	$x = 4$

Therefore, the solution to the equation is $x = \{-4, -2, 2, 4\}$.

58. D: If x is a whole number greater than 5 and the greatest common factor of x and y is greater than 1, the greatest common factor of x and y must be at least as great as y. For example, if x is 6 and y is 12, the greatest common factor is 6 and the least common multiple is 12, which is equal to y. Or if x is 6 and y is 10, the greatest common factor is 2 and the least common multiple is 30, which is greater than y. If x is 6 and y is 6, then the greatest common factor and least common multiple are both 6, making the quantities equal.

59. B: The trigonometric identity $1 + \cot^2\theta = \csc^2\theta$ can be used to rewrite $\csc^2\theta = 2\cot\theta$ as $1 + \cot^2\theta = 2\cot\theta$. Now, set the equation equal to 0 to get $\cot^2\theta - 2\cot\theta + 1 = 0$. Solve by factoring.

$$\begin{aligned} \cot^2\theta - 2\cot\theta + 1 &= 0 \\ (\cot\theta - 1)^2 &= 0 \\ \cot\theta - 1 &= 0 \\ \cot\theta &= 1 \end{aligned}$$

Thus, from the unit circle, for $0 < \theta \le 2\pi$, $\cot\theta = 1$ when $\theta = \frac{\pi}{4}$ or $\frac{5\pi}{4}$.

60. B: A simple test of each option at $x = -5$, $x = 0$, and $x = 2$ demonstrates that only choice B works:

x	**Observed**	y_a	y_b	y_c	y_d
-5	$\sim\boldsymbol{\frac{1}{6}}$	-2	$\frac{1}{6}$	$-\frac{7}{6}$	$\frac{1}{9}$
0	$\sim-\boldsymbol{\frac{1}{4}}$	$\frac{1}{2}$	$-\frac{1}{4}$	$-\frac{3}{4}$	$-\frac{1}{6}$
2	$\sim\boldsymbol{\frac{1}{6}}$	$\frac{1}{3}$	$\frac{1}{6}$	$\frac{7}{6}$	$\frac{1}{9}$

61. D: To find the probability, we count the number of evens that were rolled, divided by the number of total rolls.

$$\frac{0 + 12 + 20(2) + 18(3)}{50(3)} = \frac{106}{150} = \frac{53}{75} \approx 0.707$$

Therefore, the approximate probability of a single roll being even is 0.707.

62. C: To find the probability of rolling an even a certain number of times, use the formula:

$$({}_n\text{C}_r)p^r q^{n-r}$$

For example, to find the probability of rolling an even four times out of five rolls with a fair die:

$$({}_n\text{C}_r)p^r q^{n-r} = ({}_5\text{C}_4)(0.5)^4(0.5)^1 = \frac{5!}{1!\,4!}(0.5)^4(0.5)^1 \approx 0.156$$

To find the probability of rolling an even four times with the weighted die:

$$({}_n\text{C}_r)p^r q^{n-r} = ({}_5\text{C}_4)(0.707)^4(0.293)^1 = \frac{5!}{1!\,4!}(0.707)^4(0.293)^1 \approx 0.366$$

The calculated probabilities match those shown in choice C.

63. B: First, simplify the expression within the absolute value symbol. Recall, $i^2 = -1$.

$$\begin{aligned}(2-3i)^2-(1-4i) &= 4-12i+9i^2-1+4i \\ &= 4-12i-9-1+4i \\ &= -6-8i\end{aligned}$$

Therefore, the simplified expression is $-6-8i$.

64. C: We calculate volume of a cylinder with the following formula: $V = \pi r^2 h$, where r is the radius and h is the height. The radius is half of the given diameter, or 420 in, so $V = \pi(420)^2(220) =$ 121,857,120. We can write this in scientific notation by moving the decimal to the left 8 places: $1.22x10^8$ in^3. Answer A is the area of the base of the water tower. Answer B is the surface area of the water tower. Answer D is the volume using the value of the diameter as the radius.

65. B: If l and w represent the length and width of the enclosed area, its perimeter is equal to $2l+2w$. Since the fence is positioned x feet from the lot's edges on each side, the perimeter of the original pen is $2(l+2x)+2(w+2x)$. Since the amount of money saved by fencing the smaller area is \$648, and since the fencing material costs \$9 per linear foot, 72 fewer feet of material are used to fence around the cattle pen than would have been used to fence around the original plan. This can be expressed as the following equation.

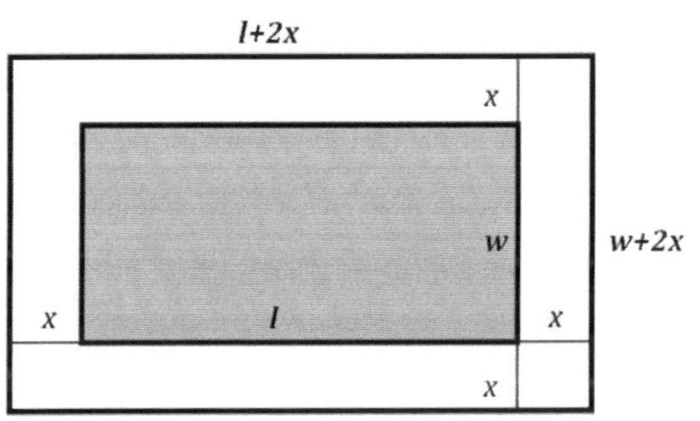

$$\begin{aligned}2(l+2x)+2(w+2x)-(2l+2w) &= 72 \\ 2l+4x+2w+4x-2l-2w &= 72 \\ 8x &= 72 \\ x &= 9 \text{ ft}\end{aligned}$$

The difference in the area of the lot and the enclosed space is 764 yd^2, which is the same as 6,876 ft^2. So, $(l+2x)(w+2x)-lw = 6{,}876$. Substituting 9 for x, and solving for $l+w$, we get the following equations.

$$\begin{aligned}(l+18)(w+18)-lw &= 6{,}876 \\ lw+18l+18w+324-lw &= 6{,}876 \\ 18l+18w &= 6{,}552 \\ 18(l+w) &= 6{,}552 \\ l+w &= 364 \text{ ft}\end{aligned}$$

Therefore, the perimeter of the enclosed space, $2(l+w)$, is $2(364) = 728$ ft. The cost of 728 ft of fencing is $728 \times \$9 = \$6{,}552$.

66. C: On the unit circle, $\sin\theta = \frac{\sqrt{3}}{2}$ when $\theta = \frac{\pi}{3}$ and when $\theta = \frac{2\pi}{3}$. Since only $\frac{2\pi}{3}$ is in the given range of $\frac{\pi}{2} < \theta < \pi, \theta = \frac{2\pi}{3}$.

67. C: The following equation may be written and solved for x: $\sin 40° = \frac{x}{6}$. Multiplying both sides of the equation by 6 gives: $6 \cdot \sin 40° = x$, or $x \approx 3.9$.

68. C: First, use the properties of logarithms to solve for y:

$2\log_8 y + \log_8 4 = 2$	
$\log_8 y^2 + \log_8 4 = 2$	Use the power rule of logarithms to rewrite the first logarithm.
$\log_8 4y^2 = 2$	Use the product rule of logarithms to combine the two logarithms.
$4y^2 = 8^2$	Rewrite the equation using the logarithm rule that $\log_a b = c$ is equivalent to $a^c = b$.
$4y^2 = 64$	Simplify the square on the right side.
$y^2 = 16$	Divide both sides by 4.
$y = 4$	Take the positive square root of both sides.

Finally, substitute 4 for y in the expression $\log_y 64$ and simplify: $\log_4 64 = 3$ since $4^3 = 64$.

69. A: First, state the exclusions of the domain.

$$x^3 + 3x^2 - 6x - 8 \neq 0$$
$$(x-2)(x+4)(x+1) \neq 0$$
$$x - 2 \neq 0 \quad x + 4 \neq 0 \quad x + 1 \neq 0$$
$$x \neq 2 \quad x \neq -4 \quad x \neq -1$$

To determine whether there are asymptotes or holes at these values of x, simplify the expression:

$$\frac{x^2 - 2x - 3}{x^3 + 3x^2 - 6x - 8} = \frac{(x-3)(x+1)}{(x-2)(x+4)(x+1)} = \frac{x-3}{(x-2)(x+4)}$$

There are asymptotes at $x = 2$ and at $x = -4$ and a hole at $x = -1$. Statement I is true.

To find the x-intercept of $f(x)$, solve $f(x) = 0$. $f(x) = 0$ when the numerator is equal to zero. The numerator equals zero when $x = 3$ and $x = -1$; however, –1 is excluded from the domain of $f(x)$, so the x-intercept is 3. To find the y-intercept of $f(x)$, find $f(0)$.

$$\frac{0^2 - 2(0) - 3}{0^3 + 3(0)^2 - 6(0) - 8} = \frac{-3}{-8} = \frac{3}{8}$$

The x-intercept is 3, but the y-intercept is not 3. Statement II is false.

70. C: To find the number of diagonals in a polygon, use the formula $\frac{n(n-3)}{2}$, where n is the number of sides on the polygon. A dodecagon has 12 sides, so substitute 12 for n and simplify.

$$\frac{12(12-3)}{2} = \frac{12(9)}{2} = \frac{108}{2} = 54$$

Therefore, a dodecagon has 54 diagonals.

71. B: If each of the three teams will contain 2 girls and 1 boy, there must be 6 girls and 3 boys in the group. The number of ways to select 2 girls from a group of 6 girls can be found using the combination formula.

$$_6C_2 = \frac{6!}{2!\,(6-2)!} = \frac{6!}{2!\,4!} = \frac{6 \times 5 \times 4!}{2!\,4!} = \frac{6 \times 5}{2 \times 1} = 15$$

The number of ways to select 1 boy from a group of 3 boys can also be found using the combination formula.

$$_3C_1 = \frac{3!}{1!\,(3-1)!} = \frac{3!}{1!\,2!} = \frac{3 \times 2!}{1!\,2!} = \frac{3}{1} = 3$$

Since each combination of girls can be paired with each combination of boys, the number of team combinations is $15 \times 3 = 45$.

72. D: Since all the answer choices are parallelograms, determine whether the parallelogram is also a rhombus, a rectangle, or both. One way to do this is by examining the parallelogram's diagonals. If the parallelogram's diagonals are perpendicular, then the parallelogram is a rhombus. If the parallelogram's diagonals are congruent, then the parallelogram is a rectangle. If a parallelogram is both a rhombus and a rectangle, then it is a square. A plot of the quadrilateral would be:

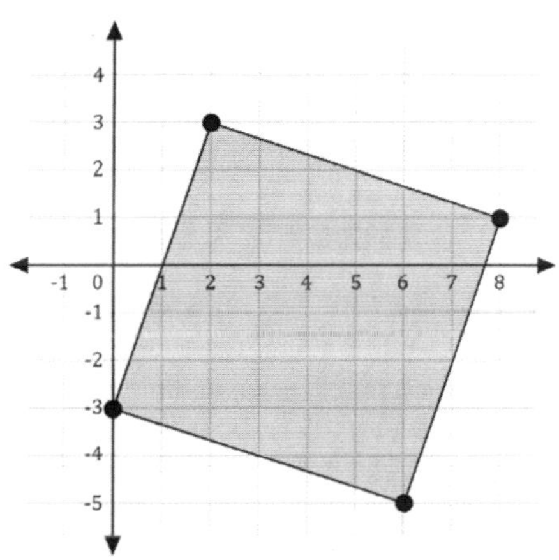

To determine whether the diagonals are perpendicular, find the slopes of the diagonals of the quadrilateral:

Diagonal 1

$$\frac{-5-3}{6-2} = \frac{-8}{4} = -2$$

Diagonal 2

$$\frac{-3-1}{0-8} = \frac{-4}{-8} = \frac{1}{2}$$

The diagonals have slopes that are opposite reciprocals of each other, so they are perpendicular. Thus, the parallelogram is a rhombus. To determine whether the diagonals are congruent, find the lengths of the diagonals of the quadrilateral:

$$\text{Diagonal 1} \quad \sqrt{(6-2)^2+(-5-3)^2} = \sqrt{(4)^2+(-8)^2} = \sqrt{16+64} = \sqrt{80}$$

$$\text{Diagonal 2} \quad \sqrt{(0-8)^2+(-3-1)^2} = \sqrt{(-8)^2+(-4)^2} = \sqrt{64+16} = \sqrt{80}$$

The diagonals are congruent, so the parallelogram is a rectangle. Since the polygon is a rhombus and a rectangle, it is also a square.

73. B: The graph of $f(x)$ is a parabola with a focus of (a, b) and a directrix of $y = -b$. The axis of symmetry of a parabola passes through the focus and vertex and is perpendicular to the directrix. Since the directrix is a horizontal line, the axis of symmetry is $x = a$; therefore, the x-coordinate of the parabola's vertex must be a. The distance between a point on the parabola and the directrix is equal to the distance between that point and the focus, so the y-coordinate of the vertex must be $y = \frac{-b+b}{2} = 0$. So, the vertex of the parabola given by $f(x)$ is $(a, 0)$.

If $g(x)$ were a translation of $f(x)$, as is the case for choices B, C, and D, the vertices of $f(x)$ and $g(x)$ would differ. Since the vertex of the graph of $g(x)$ is $(a, 0)$, none of those choices represent the correct response. However, if $g(x) = -f(x)$, the vertices of the graphs of both functions would be the same; therefore, this represents a possible relation between the two functions.

74. A: The equation of the circle is given in general form. When the equation is written in the standard form $(x-h)^2 + (y-k)^2 = r^2$, where (h, k) is the center of the circle and r is the radius of the circle, the radius can be determined by taking the positive square root of the right side. Putting the equation into standard form requires completing the square for x and y:

$$x^2 + y^2 + 2x - 10y = -18$$
$$(x^2 + 2x + 1) + (y^2 - 10y + 25) = -18 + 1 + 25$$
$$(x+1)^2 + (y-5)^2 = 8$$

Since $r^2 = 8$, and since r must be a positive number, $r = \sqrt{8} = 2\sqrt{2}$.

75. D: Set the first part of the analogy equal to y, then solve for x. When $y = \frac{1}{x}$, $x = \frac{1}{y}$. Consider choice D, when $y = x^2$, $x = \sqrt{y}$ for $y > 0$. On the other hand, when $y = x^2$, $x = \pm\sqrt{y}$; when $y = e^x$, $x = \ln y$ for $y > 0$; and when $y = x$, $x = y$.

76. A: The set of integers, $\{-3, -2, -1, 0, 1, 2, 3, \dots\}$, does not contain fractions. Since $\frac{1}{4}$ is a rational, real number, it cannot be an irrational number.

77. B: First, subtract the matrices in parentheses, then multiply the resulting matrices together.

$$\begin{bmatrix} 1 & 4 & -\frac{1}{2} \end{bmatrix} \begin{bmatrix} (2 - -2) \\ (0-5) \\ (4-2) \end{bmatrix} = \begin{bmatrix} 1 & 4 & -\frac{1}{2} \end{bmatrix} \begin{bmatrix} 4 \\ -5 \\ 2 \end{bmatrix}$$
$$= \left[(1)(4) + (4)(-5) + \left(-\frac{1}{2}\right)(2)\right]$$
$$= [4 + (-20) + (-1)]$$
$$= [-17]$$

Note that the product of a 1×3 matrix and a 3×1 matrix, in that order, is a 1×1 matrix. If the order was reversed, the product would be a 3×3 matrix.

78. C: The area of a triangle is $A = \frac{1}{2}bh$, where b and h are the lengths of the triangle's base and height, respectively. The base of the given triangle is 8, but the height is not given. Since the triangle is a right triangle and the hypotenuse is given, the triangle's height can be found using the Pythagorean theorem.

$$8^2 + h^2 = x^2$$
$$h^2 = x^2 - 8^2$$
$$h^2 = x^2 - 64$$
$$h = \sqrt{x^2 - 64}$$

To find the area of the triangle in terms of x, substitute $\sqrt{x^2 - 64}$ for the height and 8 for the base of the triangle into the area formula ($A = \frac{1}{2}bh$):

$$A(x) = \frac{1}{2}(8)(\sqrt{x^2 - 64})$$
$$A(x) = 4\sqrt{x^2 - 64}$$

79. C: First, write three equations from the information given in the problem. Since the total number of meals sold was 180, we get the equation $x + y + z = 180$. The sales generated \$1,152, so the equation becomes $7.5x + 6y + 5z = 1{,}152$. The number of students who bought pizza was half the number of the other meals put together, so $z = \frac{1}{2}(x + y)$, which can be rewritten as $x + y - 2z = 0$.

The coefficients of each equation are arranged in the rows of a 3×3 matrix, which, when multiplied by the 3×1 matrix arranging the variables x, y, and z, will give the 3×1 matrix which arranges the constants of the equations.

80. D: Bacterial growth is exponential. Let x be the number of times the population doubles, a be the number of bacteria in the colony originally transferred into the broth, and y be the number of bacteria in the broth after x doubling times. After 1 hour, the population would have doubled three times:

$$a(2)^x = y$$
$$a(2^3) = 8 \times 10^6$$
$$8a = 8 \times 10^6$$
$$a = 10^6$$

So, the number of bacteria originally transferred into the petri dish was 10^6. The equation for determining the number of bacteria is $y = (2^x) \times 10^6$. Since the bacteria double every twenty minutes, they go through three doubling times every hour. So, when the bacteria are allowed to grow for eight hours, they will have gone through $8 \times 3 = 24$ doubling times. When $x = 24$, $y = (2^{24}) \times 10^6 = 16{,}777{,}216 \times 10^6$, which is approximately 1.7×10^{13}.

81. A: The total number of students taking electives is 200, as seen in the chart on the right, and 16% of these are taking welding, so the number of students taking welding is $0.16 \times 200 = 32$. The number of male sophomores taking an elective is 20. The difference in the number of male and female students taking an elective is $108 - 92 = 16$. The difference in the number of juniors and seniors taking an elective is $83 - 59 = 24$. The greatest quantity of these is 32, the number of students taking welding.

82. C: The percentage of students taking Java or Creative Writing is $25\% + 21\% = 46\%$. The percentage of students in electives who are male is $\frac{92}{200} \times 100\% = 46\%$. So, the two quantities are equal.

83. A: 20 of the 38 sophomores taking an elective are male. The probability that a male is chosen is $\frac{\text{number of male sophomores}}{\text{number of sophomores}} = \frac{20}{38} \approx 0.53$.

84. A: The sum of two vectors is equal to the sum of their components. Using component-wise addition, we get the following vector.

$$v + w = \big((4 - 3), (3 + 4)\big) = (1,7)$$

To multiply a vector by a scalar, multiply each component by that scalar. Using component-wise scalar multiplication, we get the following vector.

$$2(1,7) = (2 \times 1, 2 \times 7) = (2, 14)$$

85. B: Choice B does not simulate random selection because the probability of different outcomes is not equal. For instance, there is only one combination of dice outcomes that can produce the value 2 or 12, whereas there are several combinations that produce the value of 7.

86. C: If x is a prime number and the greatest common factor of x and y is greater than 1, the greatest common factor of x and y must be x. The least common multiple of two numbers is equal to the product of those numbers divided by their greatest common factor. So, the least common multiple of x and y is $\frac{xy}{x} = y$. Therefore, the two quantities are the same.

87. A: There are three ways in which two men from the same class can be selected: two men can be selected from the first class, or two men can be selected from the second class, or two men can be selected from the third class.

Class 1	**Class 2**	**Class 3**
$\frac{13}{93} \times \frac{12}{92} = \frac{156}{8{,}556}$	$\frac{14}{93} \times \frac{13}{92} = \frac{182}{8{,}556}$	$\frac{19}{93} \times \frac{18}{92} = \frac{342}{8{,}556}$

Since any of these is a distinct possible outcome, the probability that two men will be selected from the same class is the sum of these outcomes.

$$\frac{156}{8{,}556} + \frac{182}{8{,}556} + \frac{342}{8{,}556} = \frac{680}{8{,}556} \approx 0.079 = 7.9\%$$

The approximate probability that two men from the same class will be chosen is 7.9%.

88. A: The critical points of the graph occur when $f'(x) = 0$.

$f(x) = 4x^3 - x^2 - 4x + 2$	
$f'(x) = 12x^2 - 2x - 4$	Take the derivative of $f(x)$.
$f'(x) = 2(6x^2 - x - 2)$	Factor a 2 out of each term in $f'(x)$.
$f'(x) = 2(3x - 2)(2x + 1)$	Factor the new polynomial in $f'(x)$.
$0 = 2(3x - 2)(2x + 1)$	Set $f'(x)$ equal to 0.
$3x - 2 = 0$ $x = \frac{2}{3}$	Solve the first factor to find the value of x that makes it 0.
$2x + 1 = 0$ $x = -\frac{1}{2}$	Solve the second factor to find the value of x that makes it 0.

The critical points of the graph are at $x = \frac{2}{3}$ and $x = -\frac{1}{2}$.

To determine if these points are relative maximums or minimums, test points on each end of the intervals surrounding them. Start by testing a point less than $-\frac{1}{2}$. For example, solve for the value of $f'(x)$ when $x = -1$.

$$f'(-1) = 12(-1)^2 - 2(-1) - 4 = 12 + 2 - 4 = 10$$

Now, test a point greater than $-\frac{1}{2}$ and less than $\frac{2}{3}$, like 0.

$$f'(0) = 12(0)^2 - 2(0) - 4 = 0 - 0 - 4 = -4$$

Therefore, when $x < -\frac{1}{2}$, $f'(x) > 0$, and when $-\frac{1}{2} < x < \frac{2}{3}$, $f'(x) < 0$, so the point where $x = -\frac{1}{2}$ is a relative maximum of the function. Now, test a point greater than $\frac{2}{3}$, like 1.

$$f'(1) = 12(1)^2 - 2(1) - 4 = 12 - 2 - 4 = 6$$

When $x > \frac{2}{3}$, $f'(x) > 0$, so the point where $x = \frac{2}{3}$ is a relative minimum. To find the corresponding y-values for these points, substitute each value of x into the original function and solve for y.

If $f''(x) > 0$ for all x in an interval, the graph of the function is concave upward on that interval, and if $f''(x) < 0$ for all x in an interval, the graph of the function is concave downward on that interval. Find the second derivative of the function and determine the intervals in which $f''(x)$ is less than zero and greater than zero.

$$f''(x) = 24x - 2$$

$$24x - 2 < 0 \qquad 24x - 2 > 0$$

$$x < \frac{1}{12} \qquad x > \frac{1}{12}$$

The graph of f is concave downward on the interval $\left(-\infty, \frac{1}{12}\right)$ and concave upward on the interval $\left(\frac{1}{12}, \infty\right)$.

89. A: The graph below shows the height h in inches of the weight on the spring above the table as a function of time t in seconds. Notice that the height is 3 inches above the table at time 0 since the weight was pulled down 2 inches from its starting position 5 inches above the table. The spring fluctuates 2 inches above and below its equilibrium point, so its maximum height is 7 inches above the table. The graph represents a cosine curve which has been inverted, stretched vertically by a factor of 2, and shifted up 5 units. The graph has also been compressed horizontally, with a period of 1 rather than 2π. So, the height of the weight on the spring as a function of time is $h = -2\cos(2\pi t) + 5$.

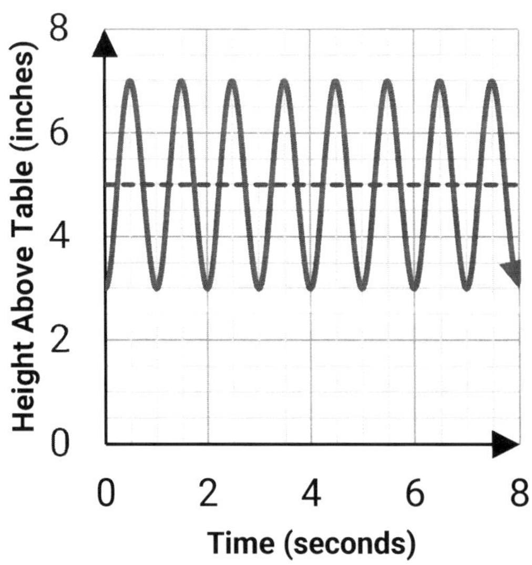

90. A: Besides the \$600 he has remaining, Chan has paid out a total of 85% (30% + 30% + 25%) of his bonus for the expenses described in the question. Therefore, the \$600 represents the remaining 15%. Remember that 15% can be written as $\frac{15}{100}$. To determine his total bonus, solve $\frac{15}{100}x = 600$. So, $x = \frac{100}{15} \times 600 = 4{,}000$, and Chan's total bonus is \$4,000.

91. D: To simplify, get as many terms as possible out from the radical as possible. Then, rationalize the denominator to remove any radicals from the denominator.

$$\begin{aligned}\sqrt{\frac{-54x^5}{80y^6}} &= \sqrt{\frac{-9x^4 \times (6x)}{16y^6 \times (5)}} \\ &= \sqrt{\frac{-(3x^2)^2 \times (6x)}{(4y^3)^2 \times (5)}} \\ &= \frac{3x^2 i\sqrt{6x}}{4y^3\sqrt{5}} \times \frac{\sqrt{5}}{\sqrt{5}} \\ &= \frac{3x^2 i\sqrt{30x}}{20y^3}\end{aligned}$$

92. D: Use the formula for the volume of a sphere to find the radius of the sphere:

$$V = \frac{4}{3}\pi r^3$$
$$288\pi = \frac{4}{3}\pi r^3$$
$$288 = \frac{4}{3}r^3$$
$$216 = r^3$$
$$6 = r$$

Then, substitute the point $(h, k, l) = (-1, 1, 0)$ and the radius $r = 6$ into the equation of a sphere:

$$(x-h)^2 + (y-k)^2 + (z-l)^2 = r^2$$
$$(x+1)^2 + (y-1)^2 + z^2 = 6^2$$
$$(x+1)^2 + (y-1)^2 + z^2 = 36$$
$$x^2 + 2x + 1 + y^2 - 2y + 1 + z^2 = 36$$
$$x^2 + y^2 + z^2 + 2x - 2y = 34$$

93. C: Since rate in mph $= \frac{\text{distance in miles}}{\text{time in hours}}$, Tessa's speed on the way to the lake and back home in mph can be expressed as $\frac{d}{1.25}$ and $\frac{d}{0.75}$, respectively, when d is the distance between Tessa's house and the lake. Since she rode 4 mph faster on her way home:

$$\frac{d}{0.75} - \frac{d}{1.25} = 4$$
$$\frac{d}{0.75} \times \frac{4}{4} - \frac{d}{1.25} \times \frac{4}{4} = 4$$
$$\frac{4d}{3} - \frac{4d}{5} = 4$$
$$15\left(\frac{4d}{3} - \frac{4d}{5}\right) = 4(15)$$
$$20d - 12d = 60$$
$$d = 7.5$$

Since the distance between Tessa's house and the lake is 7.5 miles, she rode a total distance of 15 miles in 2 hours (1 hour 15 minutes + 45 minutes = 2 hours). Therefore, her average speed was $\frac{15 \text{ miles}}{2 \text{ hours}} = 7.5$ mph.

94. D: If the distance between the two houses is 10 cm on the map, then the actual distance between the houses is 100 m. To find x, use the Pythagorean theorem.

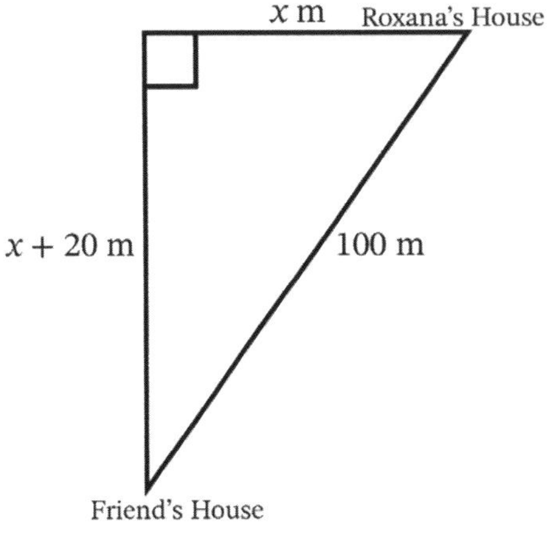

$$x^2 + (x+20)^2 = (100)^2$$
$$x^2 + x^2 + 40x + 400 = 10{,}000$$
$$2x^2 + 40x - 9{,}600 = 0$$
$$2(x^2 + 20x - 4{,}800) = 0$$
$$2(x-60)(x+80) = 0$$
$$x = 60 \quad x = -80$$

Since x represents a distance, it cannot equal -80. Since $x = 60$, $x + 20 = 80$. Roxana walks a total of $60 + 80 = 140$ meters to get to her friend's house.

95. A: A pie chart shows the relationship of parts to a whole. A line graph is often used to show change over time. A box-and-whisker plot displays how numeric data are distributed throughout the range. A Venn diagram shows the relationships among sets. Therefore, a pie chart would best illustrate the percentage of a budget allocated to various departments.

96. B: Rate in miles per hour can be expressed as, mph $= \frac{\text{distance in miles}}{\text{time in hours}}$. So, Zeke's driving speed on the way to Atlanta and home from Atlanta in mph can be expressed as $\frac{d}{3}$ and $\frac{d}{2}$, respectively, where d is the distance between Zeke's house and his destination. Since Zeke drove 20 mph faster on his way home, (speed home) − (speed to store) = 20. Substitute Zeke's speeds and solve for d.

$$\frac{d}{2} - \frac{d}{3} = 20$$

$$6\left(\frac{d}{2} - \frac{d}{3} = 20\right)$$
$$3d - 2d = 120$$
$$d = 120$$

Since the distance between Zeke's house and the store in Atlanta is 120 miles, Zeke drove a total distance of 240 miles in five hours. Therefore, his average speed was $\frac{240 \text{ miles}}{5 \text{ hours}} = 48$ mph.

97. D: If the statement is written in the form $p \rightarrow q$, then the converse is represented as $q \rightarrow p$. Thus, the converse should read, "If it is spring, I see a robin."

98. B: Percent error $= \frac{|\text{actual value} - \text{measured value}|}{\text{actual value}} \times 100\%$, and the average percent error is the sum of the percent errors for each trial divided by the number of trials.

	Percent Error Trial 1	Percent Error Trial 2	Percent Error Trial 3	Percent Error Trial 4	Average Percent Error
Scale 1	0.1%	0.2%	0.2%	0.1%	0.15%
Scale 2	2.06%	2.09%	2.10%	2.08%	2.08%

The percent error for Scale 1 is less than the percent error for Scale 2, so it is more accurate. The more precise scale is Scale 2 because its range of values, $10.210\text{ g} - 10.206\text{ g} = 0.004\text{ g}$, is smaller than the Scale 2's range of values, $10.02\text{ g} - 9.98\text{ g} = 0.04\text{ g}$.

99. A: The probability of an event is the number of possible occurrences of that event divided by the number of all possible outcomes. A student who is eight years old or less can be five, six, seven, or eight years old.

$$\frac{\text{number of 5-, 6-, 7-, and 8-year-old students}}{\text{total number of students}} = \frac{11 + 8 + 9 + 14}{11 + 8 + 9 + 14 + 11 + 13} = \frac{42}{66} = \frac{7}{11}$$

Therefore, the probability that he or she is no more than eight years old is $\frac{7}{11}$.

100. C: If $y = a(x + b)^2(x + c)^2$, the degree of the polynomial is 4. Since the degree of the polynomial is even and the leading coefficient is positive $(a > 0)$, the end behavior of the graph goes to (∞, ∞) and $(-\infty, \infty)$. Therefore, only Choice C is a possible graph of this function.

How to Overcome Test Anxiety

Just the thought of taking a test is enough to make most people a little nervous. A test is an important event that can have a long-term impact on your future, so it's important to take it seriously and it's natural to feel anxious about performing well. But just because anxiety is normal, that doesn't mean that it's helpful in test taking, or that you should simply accept it as part of your life. Anxiety can have a variety of effects. These effects can be mild, like making you feel slightly nervous, or severe, like blocking your ability to focus or remember even a simple detail.

If you experience test anxiety—whether severe or mild—it's important to know how to beat it. To discover this, first you need to understand what causes test anxiety.

Causes of Test Anxiety

While we often think of anxiety as an uncontrollable emotional state, it can actually be caused by simple, practical things. One of the most common causes of test anxiety is that a person does not feel adequately prepared for their test. This feeling can be the result of many different issues such as poor study habits or lack of organization, but the most common culprit is time management. Starting to study too late, failing to organize your study time to cover all of the material, or being distracted while you study will mean that you're not well prepared for the test. This may lead to cramming the night before, which will cause you to be physically and mentally exhausted for the test. Poor time management also contributes to feelings of stress, fear, and hopelessness as you realize you are not well prepared but don't know what to do about it.

Other times, test anxiety is not related to your preparation for the test but comes from unresolved fear. This may be a past failure on a test, or poor performance on tests in general. It may come from comparing yourself to others who seem to be performing better or from the stress of living up to expectations. Anxiety may be driven by fears of the future—how failure on this test would affect your educational and career goals. These fears are often completely irrational, but they can still negatively impact your test performance.

Elements of Test Anxiety

As mentioned earlier, test anxiety is considered to be an emotional state, but it has physical and mental components as well. Sometimes you may not even realize that you are suffering from test anxiety until you notice the physical symptoms. These can include trembling hands, rapid heartbeat, sweating, nausea, and tense muscles. Extreme anxiety may lead to fainting or vomiting. Obviously, any of these symptoms can have a negative impact on testing. It is important to recognize them as soon as they begin to occur so that you can address the problem before it damages your performance.

The mental components of test anxiety include trouble focusing and inability to remember learned information. During a test, your mind is on high alert, which can help you recall information and stay focused for an extended period of time. However, anxiety interferes with your mind's natural processes, causing you to blank out, even on the questions you know well. The strain of testing during anxiety makes it difficult to stay focused, especially on a test that may take several hours. Extreme anxiety can take a huge mental toll, making it difficult not only to recall test information but even to understand the test questions or pull your thoughts together.

Effects of Test Anxiety

Test anxiety is like a disease—if left untreated, it will get progressively worse. Anxiety leads to poor performance, and this reinforces the feelings of fear and failure, which in turn lead to poor performances on subsequent tests. It can grow from a mild nervousness to a crippling condition. If allowed to progress, test anxiety can have a big impact on your schooling, and consequently on your future.

Test anxiety can spread to other parts of your life. Anxiety on tests can become anxiety in any stressful situation, and blanking on a test can turn into panicking in a job situation. But fortunately, you don't have to let anxiety rule your testing and determine your grades. There are a number of relatively simple steps you can take to move past anxiety and function normally on a test and in the rest of life.

Physical Steps for Beating Test Anxiety

While test anxiety is a serious problem, the good news is that it can be overcome. It doesn't have to control your ability to think and remember information. While it may take time, you can begin taking steps today to beat anxiety.

Just as your first hint that you may be struggling with anxiety comes from the physical symptoms, the first step to treating it is also physical. Rest is crucial for having a clear, strong mind. If you are tired, it is much easier to give in to anxiety. But if you establish good sleep habits, your body and mind will be ready to perform optimally, without the strain of exhaustion. Additionally, sleeping well helps you to retain information better, so you're more likely to recall the answers when you see the test questions.

Getting good sleep means more than going to bed on time. It's important to allow your brain time to relax. Take study breaks from time to time so it doesn't get overworked, and don't study right before bed. Take time to rest your mind before trying to rest your body, or you may find it difficult to fall asleep.

Along with sleep, other aspects of physical health are important in preparing for a test. Good nutrition is vital for good brain function. Sugary foods and drinks may give a burst of energy but this burst is followed by a crash, both physically and emotionally. Instead, fuel your body with protein and vitamin-rich foods.

Also, drink plenty of water. Dehydration can lead to headaches and exhaustion, especially if your brain is already under stress from the rigors of the test. Particularly if your test is a long one, drink water during the breaks. And if possible, take an energy-boosting snack to eat between sections.

Along with sleep and diet, a third important part of physical health is exercise. Maintaining a steady workout schedule is helpful, but even taking 5-minute study breaks to walk can help get your blood pumping faster and clear your head. Exercise also releases endorphins, which contribute to a positive feeling and can help combat test anxiety.

When you nurture your physical health, you are also contributing to your mental health. If your body is healthy, your mind is much more likely to be healthy as well. So take time to rest, nourish your body with healthy food and water, and get moving as much as possible. Taking these physical steps will make you stronger and more able to take the mental steps necessary to overcome test anxiety.

Mental Steps for Beating Test Anxiety

Working on the mental side of test anxiety can be more challenging, but as with the physical side, there are clear steps you can take to overcome it. As mentioned earlier, test anxiety often stems from lack of preparation, so the obvious solution is to prepare for the test. Effective studying may be the most important weapon you have for beating test anxiety, but you can and should employ several other mental tools to combat fear.

First, boost your confidence by reminding yourself of past success—tests or projects that you aced. If you're putting as much effort into preparing for this test as you did for those, there's no reason you should expect to fail here. Work hard to prepare; then trust your preparation.

Second, surround yourself with encouraging people. It can be helpful to find a study group, but be sure that the people you're around will encourage a positive attitude. If you spend time with others who are anxious or cynical, this will only contribute to your own anxiety. Look for others who are motivated to study hard from a desire to succeed, not from a fear of failure.

Third, reward yourself. A test is physically and mentally tiring, even without anxiety, and it can be helpful to have something to look forward to. Plan an activity following the test, regardless of the outcome, such as going to a movie or getting ice cream.

When you are taking the test, if you find yourself beginning to feel anxious, remind yourself that you know the material. Visualize successfully completing the test. Then take a few deep, relaxing breaths and return to it. Work through the questions carefully but with confidence, knowing that you are capable of succeeding.

Developing a healthy mental approach to test taking will also aid in other areas of life. Test anxiety affects more than just the actual test—it can be damaging to your mental health and even contribute to depression. It's important to beat test anxiety before it becomes a problem for more than testing.

Study Strategy

Being prepared for the test is necessary to combat anxiety, but what does being prepared look like? You may study for hours on end and still not feel prepared. What you need is a strategy for test prep. The next few pages outline our recommended steps to help you plan out and conquer the challenge of preparation.

Step 1: Scope Out the Test

Learn everything you can about the format (multiple choice, essay, etc.) and what will be on the test. Gather any study materials, course outlines, or sample exams that may be available. Not only will this help you to prepare, but knowing what to expect can help to alleviate test anxiety.

Step 2: Map Out the Material

Look through the textbook or study guide and make note of how many chapters or sections it has. Then divide these over the time you have. For example, if a book has 15 chapters and you have five days to study, you need to cover three chapters each day. Even better, if you have the time, leave an extra day at the end for overall review after you have gone through the material in depth.

If time is limited, you may need to prioritize the material. Look through it and make note of which sections you think you already have a good grasp on, and which need review. While you are studying, skim quickly through the familiar sections and take more time on the challenging parts.

Write out your plan so you don't get lost as you go. Having a written plan also helps you feel more in control of the study, so anxiety is less likely to arise from feeling overwhelmed at the amount to cover.

Step 3: Gather Your Tools

Decide what study method works best for you. Do you prefer to highlight in the book as you study and then go back over the highlighted portions? Or do you type out notes of the important information? Or is it helpful to make flashcards that you can carry with you? Assemble the pens, index cards, highlighters, post-it notes, and any other materials you may need so you won't be distracted by getting up to find things while you study.

If you're having a hard time retaining the information or organizing your notes, experiment with different methods. For example, try color-coding by subject with colored pens, highlighters, or post-it notes. If you learn better by hearing, try recording yourself reading your notes so you can listen while in the car, working out, or simply sitting at your desk. Ask a friend to quiz you from your flashcards, or try teaching someone the material to solidify it in your mind.

Step 4: Create Your Environment

It's important to avoid distractions while you study. This includes both the obvious distractions like visitors and the subtle distractions like an uncomfortable chair (or a too-comfortable couch that makes you want to fall asleep). Set up the best study environment possible: good lighting and a comfortable work area. If background music helps you focus, you may want to turn it on, but otherwise keep the room quiet. If you are using a computer to take notes, be sure you don't have any other windows open, especially applications like social media, games, or anything else that could distract you. Silence your phone and turn off notifications. Be sure to keep water close by so you stay hydrated while you study (but avoid unhealthy drinks and snacks).

Also, take into account the best time of day to study. Are you freshest first thing in the morning? Try to set aside some time then to work through the material. Is your mind clearer in the afternoon or evening? Schedule your study session then. Another method is to study at the same time of day that you will take the test, so that your brain gets used to working on the material at that time and will be ready to focus at test time.

Step 5: Study!

Once you have done all the study preparation, it's time to settle into the actual studying. Sit down, take a few moments to settle your mind so you can focus, and begin to follow your study plan. Don't give in to distractions or let yourself procrastinate. This is your time to prepare so you'll be ready to fearlessly approach the test. Make the most of the time and stay focused.

Of course, you don't want to burn out. If you study too long you may find that you're not retaining the information very well. Take regular study breaks. For example, taking five minutes out of every hour to walk briskly, breathing deeply and swinging your arms, can help your mind stay fresh.

As you get to the end of each chapter or section, it's a good idea to do a quick review. Remind yourself of what you learned and work on any difficult parts. When you feel that you've mastered the material, move on to the next part. At the end of your study session, briefly skim through your notes again.

But while review is helpful, cramming last minute is NOT. If at all possible, work ahead so that you won't need to fit all your study into the last day. Cramming overloads your brain with more information than it can process and retain, and your tired mind may struggle to recall even previously learned information when it is overwhelmed with last-minute study. Also, the urgent nature of cramming and the stress placed on your brain contribute to anxiety. You'll be more likely to go to the test feeling unprepared and having trouble thinking clearly.

So don't cram, and don't stay up late before the test, even just to review your notes at a leisurely pace. Your brain needs rest more than it needs to go over the information again. In fact, plan to finish your studies by noon or early afternoon the day before the test. Give your brain the rest of the day to relax or focus on other things, and get a good night's sleep. Then you will be fresh for the test and better able to recall what you've studied.

Step 6: Take a Practice Test

Many courses offer sample tests, either online or in the study materials. This is an excellent resource to check whether you have mastered the material, as well as to prepare for the test format and environment.

Check the test format ahead of time: the number of questions, the type (multiple choice, free response, etc.), and the time limit. Then create a plan for working through them. For example, if you have 30 minutes to take a 60-question test, your limit is 30 seconds per question. Spend less time on the questions you know well so that you can take more time on the difficult ones.

If you have time to take several practice tests, take the first one open book, with no time limit. Work through the questions at your own pace and make sure you fully understand them. Gradually work up to taking a test under test conditions: sit at a desk with all study materials put away and set a timer. Pace yourself to make sure you finish the test with time to spare and go back to check your answers if you have time.

After each test, check your answers. On the questions you missed, be sure you understand why you missed them. Did you misread the question (tests can use tricky wording)? Did you forget the information? Or was it something you hadn't learned? Go back and study any shaky areas that the practice tests reveal.

Taking these tests not only helps with your grade, but also aids in combating test anxiety. If you're already used to the test conditions, you're less likely to worry about it, and working through tests until you're scoring well gives you a confidence boost. Go through the practice tests until you feel comfortable, and then you can go into the test knowing that you're ready for it.

Test Tips

On test day, you should be confident, knowing that you've prepared well and are ready to answer the questions. But aside from preparation, there are several test day strategies you can employ to maximize your performance.

First, as stated before, get a good night's sleep the night before the test (and for several nights before that, if possible). Go into the test with a fresh, alert mind rather than staying up late to study.

Try not to change too much about your normal routine on the day of the test. It's important to eat a nutritious breakfast, but if you normally don't eat breakfast at all, consider eating just a protein bar. If you're a coffee drinker, go ahead and have your normal coffee. Just make sure you time it so that the caffeine doesn't wear off right in the middle of your test. Avoid sugary beverages, and drink enough water to stay hydrated but not so much that you need a restroom break 10 minutes into the

test. If your test isn't first thing in the morning, consider going for a walk or doing a light workout before the test to get your blood flowing.

Allow yourself enough time to get ready, and leave for the test with plenty of time to spare so you won't have the anxiety of scrambling to arrive in time. Another reason to be early is to select a good seat. It's helpful to sit away from doors and windows, which can be distracting. Find a good seat, get out your supplies, and settle your mind before the test begins.

When the test begins, start by going over the instructions carefully, even if you already know what to expect. Make sure you avoid any careless mistakes by following the directions.

Then begin working through the questions, pacing yourself as you've practiced. If you're not sure on an answer, don't spend too much time on it, and don't let it shake your confidence. Either skip it and come back later, or eliminate as many wrong answers as possible and guess among the remaining ones. Don't dwell on these questions as you continue—put them out of your mind and focus on what lies ahead.

Be sure to read all of the answer choices, even if you're sure the first one is the right answer. Sometimes you'll find a better one if you keep reading. But don't second-guess yourself if you do immediately know the answer. Your gut instinct is usually right. Don't let test anxiety rob you of the information you know.

If you have time at the end of the test (and if the test format allows), go back and review your answers. Be cautious about changing any, since your first instinct tends to be correct, but make sure you didn't misread any of the questions or accidentally mark the wrong answer choice. Look over any you skipped and make an educated guess.

At the end, leave the test feeling confident. You've done your best, so don't waste time worrying about your performance or wishing you could change anything. Instead, celebrate the successful completion of this test. And finally, use this test to learn how to deal with anxiety even better next time.

Review Video: Test Anxiety
Visit mometrix.com/academy and enter code: 100340

Important Qualification

Not all anxiety is created equal. If your test anxiety is causing major issues in your life beyond the classroom or testing center, or if you are experiencing troubling physical symptoms related to your anxiety, it may be a sign of a serious physiological or psychological condition. If this sounds like your situation, we strongly encourage you to seek professional help.

Online Resources

Due to our efforts to try to keep this book to a manageable length, we've created a link that will give you access to all of your online resources:

mometrix.com/resources719/orelamath

It's Your Moment, Let's Celebrate It!

Share your story @mometrixtestpreparation